Phys. Chem. Glasses: Eur. J. Glass Sci. Technol. B, 2025, **66**(3), 81–86

Water diffusion in lead silicate and sodium lead silicate melts

J. E. Shelby

New York State College of Ceramics, Alfred University, Alfred, NY 14802, USA

Manuscript received 4 January 2025
Revision received 21 March 2025
Manuscript accepted 26 March 2025

Water diffusion in lead silicate melts has been measured at temperatures ranging from 800 to 1200°C for compositions containing 35 to 65 mol% PbO. The diffusivity of water increases with increasing PbO concentration and with experimental temperature, with values ranging from 1×10^{-6} to 5×10^{-5} cm^2/s. Water diffusion was also measured at 1200°C for three series of sodium lead silicate melts. The diffusivity of water increases with decreasing SiO_2 content. The results of this study are compared to those for water diffusivity in sodium silicate and soda lime silicate melts. The relation between water diffusivity and melt viscosity is considered for all four glass forming systems. It is demonstrated that the correlation between water diffusivity and melt viscosity only occurs for very limited ranges of temperature and melt composition. As has been found for many other systems, the glass transformation temperatures of the lead silicate and sodium lead silicate glasses decrease with increasing water concentration in the glass.

1. Introduction

All oxide glasses contain some 'water' in the form of either hydroxyl or molecular water or both.[1] The relative concentrations of these species which are formed by reaction with atmospheric water vapour varies with melt temperature and partial pressure of water in the melting atmosphere. The ratio of molecular water to hydroxyl in melts prepared under partial pressures of one atmosphere or less is quite small.[2] When one considers that the detection of these species by infrared spectroscopy is much more sensitive to the presence of hydroxyl than to molecular water,[3] it is obvious that the 'water' content reported for glasses melted under normal ambient pressures is only that of the hydroxyl in the glass when the water content is measured by infrared spectroscopy.

If a glass is remelted under an atmosphere in which the concentration of water vapour is different from that in equilibrium with that in the initial melt or is changed during the melting process, the concentration of hydroxyl in the melt will adjust to the new equilibrium concentration.[1] During this adjustment period, a gradient in hydroxyl will occur in the melt and thus in the glass formed by cooling the melt. In the absence of stirring or convection, the time required to eliminate this gradient and to reach equilibrium will depend on the effective diffusivity of water in the melt. Since changes in hydroxyl concentration result in changes in the glass transformation temperature ranging from less than 10 K for very fragile melts to over 100 K for very strong melts[4] as well as changes of orders of magnitude in the transformation range viscosity[5] and in the rate of crystallisation,[5,6] controlling the hydroxyl concentration is essential for controlling the properties of glasses.

The effective diffusivity of water, which will be referred to as the 'water diffusivity' throughout this paper, in oxide melts is a strong function of melt composition and melting temperature.[1] Unfortunately, water diffusivity in melts of interest to glass scientists and technologists, as opposed to geologists, at water partial pressures of no more than one atmosphere has been the subject of relatively limited study. Almost all the published studies have dealt with either single component glasses (SiO_2, B_2O_3, or GeO_2) or binary systems containing only an alkali oxide and a glass former.[1,2,7–12] A few studies have dealt with the more complex compositions found in commercial silicate glasses.[13–17] There are no reported studies of binary systems involving oxides of divalent ions. Studies of ternary systems are restricted to the sodium alkaline earth silicate[18,19] and sodium aluminosilicate melts[20] and to a temperature of 1200°C. These studies have been summarised by Shelby.[21]

This paper presents the results of water diffusivity measurements in lead silicate melts containing 35 to 65 mol% PbO at temperatures ranging from 800 to 1200°C. These results are the first values reported for *R*O–SiO_2 melts. The water

Email address: shelbyje@alfred.edu
DOI: 10.13036/17533562.66.3.01

diffusivity in three series of sodium lead silicate melts was measured at 1200°C. These values are compared to the published results for the soda lime silicate melts.

2. Experimental procedure

Melts were prepared from PbO, Na_2CO_3, and SiO_2 in batches calculated to yield 15 g of lead silicate and sodium lead silicate glasses. These batches were melted in Pt alloy crucibles in an electrically heated furnace in ambient air at temperatures ranging from 1000–1400°C, depending on the batch composition. This melting procedure yielded glasses with much lower hydroxyl concentrations that those obtained during subsequent remelting in an atmosphere saturated with water vapour. Melting temperatures decreased as the PbO concentration increased. The melts were quenched by dipping the bottom of the crucible into cold water, which resulted in fragmentation of the glass.

The method for determining the water diffusivity has been explained in detail in several other papers[8,9,11,12] and will only be discussed briefly here. A small batch of the crushed glass is remelted in Pt/5Au crucibles in ambient air to yield an approximately 1·5 to 2·0 mm thick bubble-free sample. The sample is removed from the crucible, the infrared spectrum and sample thickness are measured to establish the initial values for the untreated sample, the glass is replaced in the crucible in its original orientation, and the crucible is replaced in the water treatment furnace at the desired measurement temperature. The water treatment furnace atmosphere is controlled by passing flowing air saturated at one atmosphere of water vapour (boiling distilled water) throughout the measurement. The sample is periodically removed from the furnace and cooled to room temperature. The infrared spectrum and sample thickness are measured at several points and the sample is replaced in the crucible in the original orientation. This process is continued until there are no further changes in the infrared spectrum, i.e. the sample is saturated for the specified experimental conditions.

Determination of the water diffusivity is based on a method described by Crank[22] and discussed in detail in several other publications.[1,8,9,11–20] The average value of the absorbance/thickness at the maximum in the band due to hydroxyl in the glass[23] at 3500 to 3450 cm^{-1} after a known treatment time is converted to the fraction of reaction (F) by subtracting the value for the untreated sample (A_0) from that of the treated sample (A^t) and dividing the difference by the difference between the value obtained by subtracting the value for the untreated sample (A_0) from that of the saturated sample (A^{sat}), as expressed by the equation

$$F = \frac{A_t - A_0}{A_{sat} - A_0} \tag{1}$$

The fraction of reaction is then plotted against the square root of the treatment time divided by the sample thickness. The initial slope, R, of the best fit line through the data below 50% fraction of reaction is then used to calculate the effective diffusivity, D, using the expression

$$D = \left(\pi/4\right) R^2 \tag{2}$$

The T_g was determined using a differential scanning calorimeter (DSC) with a heating rate of 20 K/min. The DSC was calibrated by measuring the melting temperatures of indium, tin, lead, and zinc and the α–β inversion temperature of quartz.

3. Results

Water diffusivities were measured for binary lead silicate (PS) melts over a wide range of temperatures and at 1200°C for ternary sodium lead silicate (NPS) melts. The results of this study are compared to those for the most closely related analogs for which data are available in the literature, i.e. the binary sodium silicate (NS) and ternary soda lime silicate (NCS) melts to clarify the effect of PbO on water diffusivity in silicate melts.

The water diffusivity in lead silicate (PS) melts increases with increasing temperature (Figure 1) and PbO concentration (Figure 2). These results are consistent with those previously reported for sodium silicate (NS) melts,[8–10] where the water diffusivity also increases with temperature and Na_2O concentration. The compositions of the lead silicate melts, along with the measured water diffusivities and the melt viscosities,[24] are listed in Table 1.

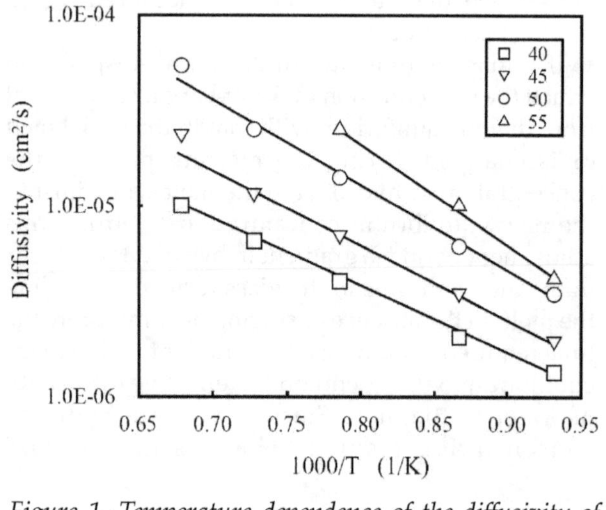

Figure 1. Temperature dependence of the diffusivity of water in lead silicate melts

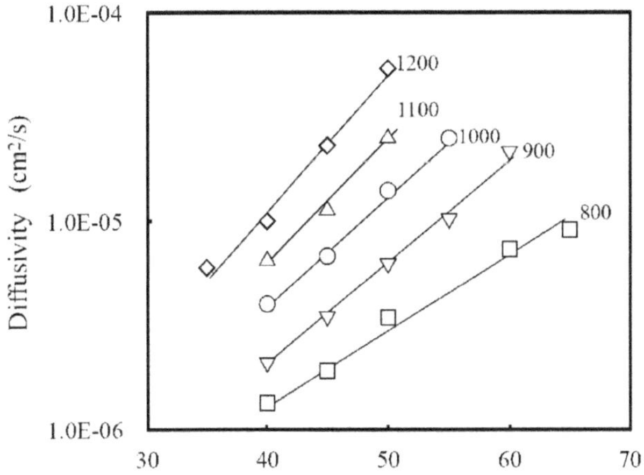

Figure 2. Compositional dependence of the diffusivity of water in lead silicate melts

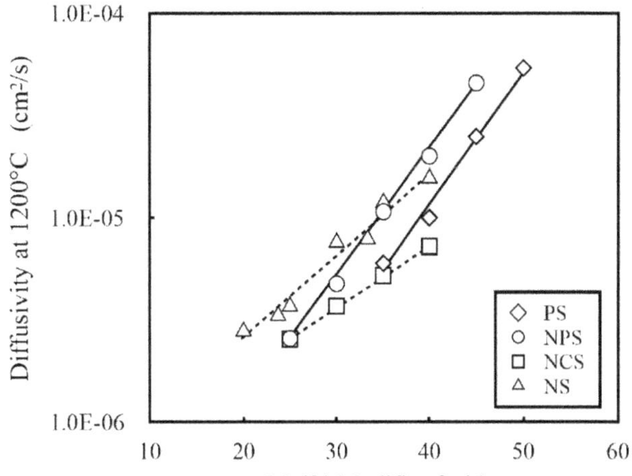

Figure 3. Comparison of the compositional dependence of water diffusivity at 1200°C in $PbO–SiO_2$ and $Na_2O–PbO–SiO_2$ melts with that in $Na_2O–SiO_2$ and $Na_2O–CaO–SiO_2$ melts from previous studies[9,18]

Results for the lead silicate melts at 1200°C are compared to those of the sodium silicate melts in Figure 3, along with those for some sodium lead silicate and soda lime silicate melts. The values taken from the literature[8,9] for the sodium silicate melts are listed in Table 2. The slope of the water diffusivity versus melt composition line for the sodium silicate melts is obviously quite different from that for the lead silicate melts.

The water diffusivity in sodium lead silicate (NPS) melts was measured at 1200°C for three compositional series: $x Na_2O–x PbO–(100–2x)SiO_2$ (equimolar ratio of Na_2O to PbO), $20Na_2O–x PbO–(80–x)SiO_2$ (constant Na_2O concentration), and $(25–x)Na_2O–x PbO–75SiO_2$ (constant SiO_2 concentration). The melt compositions, water diffusivities, and melt viscosities[25] for the sodium lead silicate melts are listed in Table 3. At this temperature, the water diffusivity in melts containing an equimolar ratio of Na_2O to PbO [$x Na_2O–x PbO–(100–2x)SiO_2$ melts] increases as the total concentration of modifier oxides increases, i.e. as the silica concentration decreases (see Figure 3). Replacement of half of the PbO by Na_2O increases the water diffusivity relative to that for a binary lead silicate melt containing the same total concentration of modifier oxides. This behaviour is the reverse of that found for replacement of Na_2O by CaO in soda lime silicate (NCS) melts containing equimolar ratios of Na_2O to CaO [$x Na_2O–x CaO–(100–2x)SiO_2$ melts], where replacement of half of the Na_2O by CaO decreases the water diffusivity.[18] The melt compositions, water diffusivities, and melt viscosities for the soda lime silicate melts are listed in Table 4.[18] The slopes of the log diffusivity versus modifier oxide concentration are similar for the lead silicate and sodium lead silicate melts. These slopes are quite different from those of the sodium silicate[8–10] and soda lime silicate[18] melts.

The effect of the replacement of SiO_2 by PbO or CaO on the water diffusivity in melts containing 20 mol% Na_2O [$20Na_2O–x RO–(80–x)SiO_2$ melts, where R is Pb or Ca] is shown in Figure 4. The initial increase in total modifier concentration, i.e. replacement of 5 mol% of SiO_2 by either PbO (squares) or CaO (circles) has little effect on the

Table 1. Compositions, measurement temperatures, water diffusivities, and viscosities[24] *of the lead silicate melts*

Mol% PbO	*Measurement temperature (°C)*	*Water diffusivity (cm²/s)*	*Viscosity (Pa-s)*
35	1200	$6·0×10^{-5}$	6·2
40	800	$1·3×10^{-6}$	182
	900	$2·0×10^{-6}$	33
	1000	$4·0×10^{-6}$	11
	1100	$6·5×10^{-6}$	4·3
	1200	$1·0×10^{-5}$	2·0
45	800	$1·9×10^{-6}$	33
	900	$3·4×10^{-6}$	8·5
	1000	$6·8×10^{-6}$	3·4
	1100	$1·1×10^{-5}$	1·2
	1200	$2·3×10^{-5}$	0·88
50	800	$3·4×10^{-6}$	7·7
	900	$6·1×10^{-6}$	2·3
	1000	$1·4×10^{-5}$	1·0
	1100	$2·5×10^{-5}$	0·54
	1200	$5·4×10^{-5}$	0·29
55	800	$4·2×10^{-6}$	1·8
	900	$1·0×10^{-5}$	0·71
	1000	$2·5×10^{-5}$	0·34
60	800	$7·4×10^{-6}$	0·47
	900	$2·1×10^{-5}$	0·22
65	800	$9·1×10^{-6}$	0·29

Table 2. Water diffusivities and viscosities at 1200°C of the sodium silicate melts[8,9]

Mol% Na_2O	*Water diffusivity (cm²/s)*	*Viscosity (Pa-s)*
20	$2·78×10^{-6}$	79·4
23·7	$3·30×10^{-6}$	50·0
25	$3·69×10^{-6}$	39·8
30	$7·54×10^{-6}$	22·4
33·3	$7·89×10^{-6}$	18·7
35	$1·22×10^{-5}$	11·2
40	$1·56×10^{-5}$	5·8

Table 3. Water diffusivities and viscosities at 1200°C of the sodium lead silicate melts

Composition	*Water diffusivity (cm^2/s)*	*Viscosity (Pa-s)*
$12{\cdot}5Na_2O–12{\cdot}5PbO–75SiO_2$	$2{\cdot}56\times10^{-6}$	-
$15Na_2O–10PbO–75SiO_2$	$3{\cdot}01\times10^{-6}$	-
$20Na_2O–5PbO–75SiO_2$	$3{\cdot}61\times10^{-6}$	-
$15Na_2O–15PbO–70SiO_2$	$4{\cdot}75\times10^{-6}$	13·8
$20Na_2O–10PbO–70SiO_2$	$7{\cdot}37\times10^{-6}$	14·5
$17{\cdot}5Na_2O–17{\cdot}5PbO–65SiO_2$	$1{\cdot}07\times10^{-5}$	5·4
$20Na_2O–15PbO–65SiO_2$	$1{\cdot}03\times10^{-5}$	5·0
$20Na_2O–20PbO–60SiO_2$	$2{\cdot}00\times10^{-5}$	3·2
$20Na_2O–25PbO–55SiO_2$	$3{\cdot}51\times10^{-5}$	0·8
$22{\cdot}5Na_2O–22{\cdot}5PbO–55SiO_2$	$4{\cdot}57\times10^{-5}$	0·9

water diffusivity. Further additions of PbO or CaO result in much larger increases in the water diffusivity. The effect of PbO is much larger than the effect of CaO on the increases in water diffusivity.

The effect of replacement of Na_2O by PbO or CaO while maintaining a constant SiO_2 concentration [$(25-x)Na_2O–xRO–75SiO_2$ melts, where R is Pb or Ca] on the water diffusivity is very different from that for the melts containing a constant Na_2O concentration, as also shown in Figure 4. The initial replacement of Na_2O by either PbO (point up triangles) or CaO (point down triangles) decreases the diffusivity of water in these melts. Further replacement of Na_2O by either PbO or CaO has little effect on the water diffusivity. There is very little difference between the effects of PbO and CaO on the water diffusivity of these melts.

The relationship between the water diffusivity in lead silicate melts and the melt viscosity is shown in Figure 5. Viscosities for the lead silicate melts are taken from Konovapov & Evstrop'iev.[24] The water diffusivity at each measurement temperature increases as the viscosity of the melt decreases, i.e. as the PbO concentration increases. A similar effect occurs for sodium silicate melts.[8–10] The water diffusivity at a given viscosity also increases as the melt temperature increases for the lead silicate melts. This is the reverse of the effect observed for sodium silicate melts, where the water diffusivity at a given viscosity decreases as the melt temperature increases.[9] It is obvious that the water diffusivity of lead silicate melts does not depend directly on the

Table 4. Water diffusivities and viscosities at 1200°C of the soda lime silicate melts[18]

Composition	*Water diffusivity (cm^2/s)*	*Viscosity (Pa-s)*
$10Na_2O–15CaO–75SiO_2$	$2{\cdot}48\times10^{-6}$	110
$12{\cdot}5Na_2O–12{\cdot}5CaO–75SiO_2$	$2{\cdot}54\times10^{-6}$	89·1
$15Na_2O–10CaO–75SiO_2$	$2{\cdot}68\times10^{-6}$	68·4
$20Na_2O–5CaO–75SiO_2$	$2{\cdot}95\times10^{-6}$	50·1
$15Na_2O–15CaO–70SiO_2$	$3{\cdot}69\times10^{-6}$	33·1
$20Na_2O–10CaO–70SiO_2$	$4{\cdot}02\times10^{-6}$	25·7
$25Na_2O–5CaO–70SiO_2$	$5{\cdot}54\times10^{-6}$	20·4
$17{\cdot}5Na_2O–17{\cdot}5CaO–65SiO_2$	$5{\cdot}22\times10^{-6}$	23·4
$20Na_2O–15CaO–65SiO_2$	$4{\cdot}77\times10^{-6}$	-
$25Na_2O–10CaO–65SiO_2$	$5{\cdot}84\times10^{-6}$	21·4
$20Na_2O–20CaO–60SiO_2$	$7{\cdot}24\times10^{-6}$	15·1
$25Na_2O–15CaO–60SiO_2$	$8{\cdot}05\times10^{-6}$	12·0

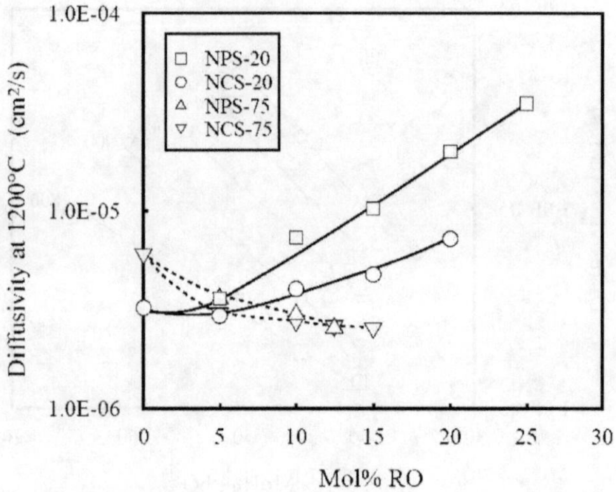

Figure 4. Comparison of water diffusivity at 1200°C in $Na_2O–PbO–SiO_2$ and $Na_2O–CaO–SiO_2$ melts containing either 20 mol% Na_2O or $75SiO_2$[18]

melt viscosity.

The relation between the water diffusivities and the melt viscosities at 1200°C also depend on the components of the melt, as shown in Figure 6. Partial replacement of PbO by Na_2O (sodium lead silicate melts) results in an increase in the water diffusivity as the viscosity decreases. Viscosities of the sodium lead silicate melts are taken from Skornyakov.[25] The water diffusivities of the sodium lead silicate melts are greater than those of the lead silicate melts at a given viscosity. Replacement of PbO by CaO results in a further increase in the water diffusivities at any given viscosity. The highest water diffusivities at a given viscosity occur for the binary sodium silicate melts. Partial replacement of Na_2O by CaO (soda lime silicate melts) results in a decrease in the water diffusivity at a given viscosity,[18] which is the opposite of the effect of partial replacement of PbO by Na_2O observed in this study.

It has repeatedly been shown that increasing the

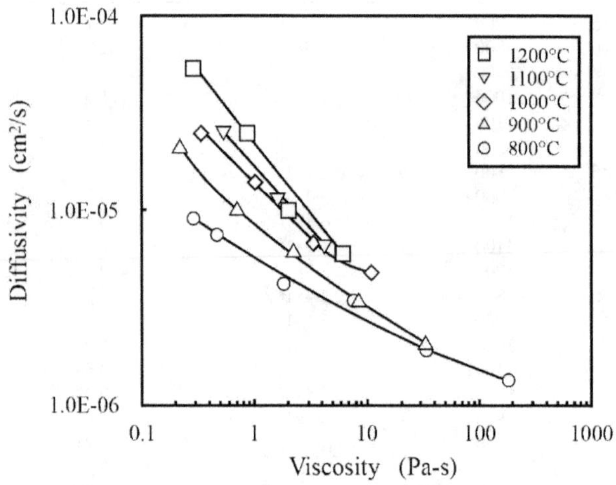

Figure 5. Relation between water diffusivity and melt viscosity in lead silicate melts

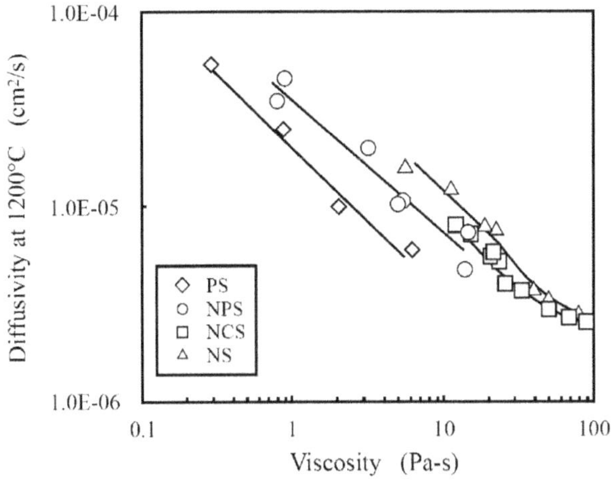

Figure 6. Relation between water diffusivity at 1200°C and melt viscosity in $PbO–SiO_2$, $Na_2O–PbO–SiO_2$, $Na_2O–SiO_2$, and $Na_2O–CaO–SiO_2$ melts

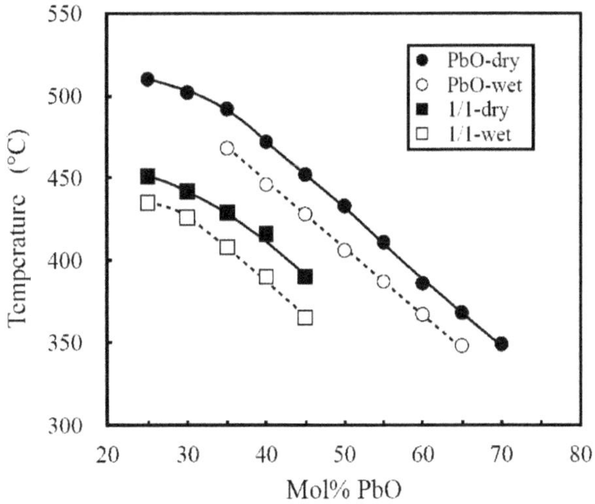

Figure 7. Effect of saturation of melts with H_2O on the glass transformation temperature of $xPbO–(100–x)SiO_2$ and $xNa_2O–(xPbO–((100–2x)SiO_2$ glasses

hydroxyl concentration of glasses decreases their glass transformation temperature.(1,4,9,11,12,26–31) The lead silicate and sodium lead silicate glasses exhibit the same behaviour as all other oxide glasses, as is shown in Figure 7. The glass transformation temperatures were measured for the glasses melted in laboratory air (dry) and after saturation at 1 atm of water vapour (wet). The values for the wet lead silicate glasses are the average of the values for the glasses saturated at each treatment temperature. There was no systematic effect of saturation temperature on the glass transformation temperature of the wet glasses, with all the saturated samples for a given PbO concentration yielding values for the glass transformation temperature within ±3 K. Since the sodium lead silicate melts were only treated at 1200°C, the wet values were obtained for those melts.

4. Discussion

The mechanism underlying water diffusion in oxide glasses and melts is difficult to determine due to the effect of changes in experimental conditions on the measured diffusivity values. For example, water diffusivity is a strong function of the water content of the melt. As the partial pressure above a melt increases, the dissolved 'water' content shifts from primarily in the form of hydroxyl to primarily in the form of water molecules.(2) The viscosity of the melts also changes significantly with total water content.(1) Water diffusivity values measured for glasses below the glass transformation range indicate that the activation energy for water diffusion in a glass and its melt are approximately the same, which negates arguments involving the viscosity of the melt. Water diffusivity has also been found to be different for the entry and removal of water.

Water diffusivity is also very sensitive to melt composition. It one compares the effect of melt composition on the water diffusivity at 1200°C, it is obvious that the components present play a large role in determining the water diffusivity. The diffusivity of water in sodium silicate melts at 1200°C, for example, varies by a factor of about two for a 10 mol% change in Na_2O concentration.(9) The same change Na_2O concentration for sodium aluminosilicate melts results in a change in water diffusivity of a factor of 7·0. A change in PbO concentration of 10 mol% in lead silicate melts, as reported in this paper, results in change in water diffusivity of a factor of five.

It is probable that the mechanism underlying water diffusion differs with both the partial pressure of atmospheric water during the experiment and with the concentration of dissolved water in the initial sample. Behrens & Nowak(32) reviewed several possible mechanisms for water diffusion in silicate melts in detail. These mechanisms include the coupled diffusion of H^+ and OH^- ions, the coupled migration of alkali ions and OH^- ions, OH diffusion by exchange of bonds on adjacent silicon–oxygen tetrahedra, migration of water molecules, and the interconversion-diffusion mechanism, which involves movement of water molecules and their reaction with bridging oxygens. They conclude that the last model is most likely for melts such as those of aluminosilicate minerals. A more recent review by Zhang & Gan(33) contends that water diffusion in melts is primarily due to the migration of water molecules and to interconversion of OH^- and water molecules at low total water contents. They also state that the mechanism(s) for water diffusion in melts will likely continue to be refined as more data becomes available. The diffusion of water molecules, with some contribution of the interconversion of OH^- and H_2O

molecules appears to adequately explain the results of the present study.

The concept that water diffusion in alkali silicate melts is controlled by the melt viscosity was originally proposed by Schulze & Mulfinger.(10) A plot of their data for sodium silicate melts containing 20·2, 24·9, or 27·2 mol% Na_2O at temperatures ranging from 1000 to 1400°C appears to support this contention, with all their data lying on a single line on a plot of log water diffusivity versus log viscosity. A later study by Mesko *et al*,(9) however, clearly shows that their results over a much broader compositional range (18 to 40 mol% Na_2O) at temperatures ranging from 900 to 1200°C do not agree with this model. The results of Mesko *et al*(9) at a given temperature do lie on a single line on a log water diffusivity versus log viscosity plot. However, this line shifts systematically with measurement temperature, with a decrease in water diffusivity at a given viscosity as the measurement temperature increases. The results reported here for the lead silicate melts also vary with melting temperature, as shown in Figure 5, but the shift with increasing temperature is toward an increase in water diffusivity at a given viscosity.

The diffusivity of water in melts always increases with increasing temperature, as does the fluidity of the melt, i.e. the viscosity decreases with increasing temperature. While both are controlled by changes in the melt structure,(1) it does not follow that the diffusivity of water must be entirely controlled by the fluidity of the melt. While the models for water diffusion discussed above indicate that the melt fluidity is a factor in determining the water diffusivity, they also contend that other factors such as the kinetics of the interconversion of OH^- and H_2O molecules may also contribute to the diffusion process.

5. Conclusions

Water diffusivity in lead silicate and sodium lead silicate melts increases with increasing PbO concentration and melt temperature. The effect of replacing Na_2O by PbO on water diffusivity is the reverse of the effect of replacing Na_2O by CaO in silicate melts. Water diffusivity in these melts is not controlled solely by the melt viscosity. The diffusivity of water in oxide melts is adequately explained by the diffusion of water molecules, with some possible contribution from the interconversion of OH^- and H_2O molecules.

References

1. Shelby, J. E. *Introduction to Glass Science and Technology, 3rd Ed.*, Royal Society of Chemistry, London, 2021.
2. Doremus, R. H. *Diffusion of Reactive Molecules in Solids and Melts*, John Wiley & Sons, Inc., New York, 2002.
3. Stolper, E., *Contributions to Mineralogy and Petrology*, 1982, **81**, 1–17.
4. Jewell, J. M., Shaw, C. M. & Shelby, J. E. *J. Non-Cryst. Solids*, 1993, **152**, 32–41.
5. Bruckner, R. *J. Non-Cryst. Solids*, 1971, **5**, 177–226.
6. Heslin, M. R. & Shelby, J. E. *Phys. Chem. Glasses*, 2003, **44**(6), 434–9.
7. McGinnis, P. B. & Shelby, J. E., *J. Non-Cryst. Solids*, 1994, **179**, 185–93.
8. Mesko, M. G. & Shelby, J. E. *Phys. Chem. Glasses*, 2001, **42**(3), 173–8.
9. Mesko, M. G., Schader, P. A. & Shelby, J. E. *Phys. Chem. Glasses*, 2002, **43**(6), 283–90.
10. Scholze, H. & Mulfinger, H. O. *Glastech. Ber.*, 1959, **32**, 381–406.
11. Shelby, J. E. *Phys. Chem. Glasses*, 2003, **44**(2), 106–12.
12. Hall, M. M. & Shelby, J. E. *Phys. Chem. Glasses*, 2004, **45**(5), 283–90.
13. Mesko, M. G. & Shelby, J. E. *Glastechn. Ber. Glass Sci. Technol.*, 2000, **73**, 13–22.
14. McGinnis, P. B. & Shelby, J. E., *J. Non-Cryst. Solids*, 1994, **177**, 381–8.
15. Mesko, M. G. & Shelby, J. E., *Phys. Chem. Glasses*, 2001, **42**(1), 17–22.
16. Mesko, M. G. & Shelby, J. E., *Phys. Chem. Glasses*, 2001, **42**(6), 389–96.
17. McGinnis, P. B., Mesko, M. G., Rapp, D. B. & Shelby, J. E., *Ceram. Trans.*, 2004, **141**, 127–34.
18. Rapp, D. B. & Shelby, J. E. *Phys. Chem. Glasses*, 2003, **44**(6), 393–400.
19. Rapp, D. B. & Shelby, J. E. *Phys. Chem. Glasses: Eur. J. Glass Sci. Technol. B*, 2021, **62**(4), 113–20.
20. Rapp, D. B. & Shelby, J. E., *Phys. Chem. Glasses: Eur. J. Glass Sci. Technol. B*, 2022, **63**(5), 129–35.
21. Shelby, J. E. *J. Am. Ceram. Soc.*, 2008, **91**, 703–8.
22. Crank, J., *The Mathematics of Diffusion*. Second Edition, Clarendon Press, Oxford, 1975.
23. Florence, J. M., Glaze, F. W. & Black, M. H. *J. Res. Nat. Bur. Standards*, 1955, **55**, 231–7.
24. Konovapov, A. S. & Evstrop'iev, K. S. Cited in *Properties of Glasses and Glass-Forming Melts.* Edited by O. V. Mazurin, M. V. Strel'tsina & T. P. Shvaiko-Shvaikoskaya, Izdatel'stvo Nauka, Leningrad, 1973, **Vol. 1**.
25. Skornyakov, M. M. Cited in *Properties of Glasses and Glass-Forming Melts.* Edited by O. V. Mazurin, M. V. Strel'tsina & T. P. Shvaiko-Shvaikoskaya, Izdatel'stvo Nauka, Leningrad, 1977, **Vol. 3**.
26. Shelby, J. E. & McVay, G. L. *J. Non-Cryst. Solids*, 1976, **20**, 439–49.
27. Jewell, J. M. & Shelby, J. E., *J. Non-Cryst. Solids*, 1988, **102**, 24–9.
28. Jewell, J. M., Spess, M. S. & Shelby, J. E. *J. Am. Ceram. Soc.*, 1990, **73**, 132–5.
29. Jewell, J. M. & Shelby, J. E. *J. Am. Ceram. Soc.*, 1992, **75**, 878–83.
30. Visser, T. J. M. & Stevels, J. M. *J. Non-Cryst. Solids*, 1972, **7**, 401.
31. Bruckner, R., *Glastech. Ber.*, 1964, **37**, 413.
32. Behrens, H. & Nowak, M., *Contrib. Mineral Petrol.*, 1997, **126**, 377–85.
33. Zhang, Y. & Gan, T. *Rev. Mineral. Geochem.*, 2022, **87**, 283–337.

Phys. Chem. Glasses: Eur. J. Glass Sci. Technol. B, 2025, **66**(3), 87–96

Mechanical and thermophysical properties of $La_2O_3.Y_2O_3.Al_2O_3$ glasses and melts

Zhitao Shan,[a] Shujiang Liu,[a] Yanfei Zhang,[a] Jiayan Zhang,[a] Qiuju Zheng,[a] Xiujun Han[a] & Haizheng Tao[b*]*

[a] *School of Materials Science and Engineering, Qilu University of Technology, Jinan 250353, China.*
[b] *State Key Laboratory of Silicate Materials for Architectures (Wuhan University of Technology), Wuhan 430070, China*

Manuscript received 24 March 2024
Revised version received 30 December 2024
Accepted 11 March 2025

Over the past few decades, rare earth aluminate glasses have attracted widespread attention in different fields, owing to remarkable electric insulation performance, high refractive index and high transmittance. In this work, we firstly investigate the mixed rare earth effect by partial substitution of lanthanum for yttrium oxides in aluminate glasses by a contactless aerodynamic laser heated levitation technique. Different with previous reported mixed modifier glasses, the result of the aluminate glasses manifest a near-linear trend in Vickers hardness with the rare earth substitution. Using Raman spectroscopies to insights into the structural these glasses, we demonstrate that the high crack resistance (CR) of aluminate glass is stronger correlation with the medium-range structure. In addition, we found the mixed rare earth effects on the glass transition temperatures, which show is stronger correlation with the medium-range structure order (peak width at half height of the lower-frequency region). In the aluminate glasses, the mixed rare earth have the Al–O and La–O bond distances diversity, which consequently increase the coefficient thermal expansion in aluminate melts. However, the isokom temperatures in aluminate melts exhibit a linear trend with composition.

1. Introduction

The 'mixed modifier effect' (MME), i.e. the non-additive behaviour of certain properties with the substitution of one modifier by another keeping the total modifier concentration constant, is one of the most intriguing unsolved issue in glass field.[(1,2)] Specifically, glasses containing mixed modifiers exhibit non-additive variations in transport properties (electrical conductivity, hardness, diffusion or viscosity) and relatively small deviations from linearity in static properties (density, refractive index, molar volume).[(3–5)] This phenomenon has been observed and investigated in many different oxide glass systems, e.g. silicate, germinate, borate, tellurite, aluminate, aluminosilicate and phosphate glasses.[(4–10)] To explore the microscopic origin of the MME on Vickers hardness (H_V), Kjeldsen *et al*[(11)] pointed out that this effect was attributed to an increase in the propensity for shear flow deformation upon loading in mixed composition. Considering the indentation–loading behaviour, Shan *et al* found that the mixed alkaline earth effect on H_V is much weaker in aluminate glasses than that in silicate glasses since the alkaline earth ions act as a charge balancer rather than modifier in the network structure of aluminate glass.[(7)]

Different from the mixed modifier effect, the mixed rare-earth effect in oxide glasses, e.g. aluminate glasses, has not been studied until now. Rare earth aluminate glass is also a kind of functional material with excellent optical and magnetic properties, which can be used as sensors, laser and radiation shielding glasses.[(12)] The rare-earth aluminate binary system displays a diversity of nonequilibrium phenomena, including crystallisation, glass formation, and polyamorphism in the undercooled liquid.[(13)] Highly crystalline materials in the Y_2O_3–Al_2O_3 system have been intensely studied because orthorhombic $YAlO_3$ and cubic $Y_3Al_5O_{12}$ have been of widely used as solid state laser materials.[(14)] However, the process by which undercooled YAG solidifies is complex and tends to generate metastable solids and glasses that do not appear in the equilibrium phase diagram of the Y_2O_3–Al_2O_3 system. It is reported that the $Y_3Al_5O_{12}$ glasses and compounds in the Y_2O_3–Al_2O_3 system contain two glass phases with identical chemical compositions but different structures.[(15–18)] Hence, it is necessary to investigate the thermophysical properties of melts to understand the complex solidification mechanism occurring in melts. In binary system, the glass forming region increased with increasing ionic radius of the rare

[1] Corresponding authors. Email jia_yan_zhang@126.com and thz@whut.edu.cn
DOI: 10.13036/17533562.66.3.12

earth element. Among different matrices, La_2O_3–Al_2O_3 glasses play an indispensable role because of good glass forming ability, wide transmission region, low cost, together with thermal and chemical durability.[19]

However, preparation of bulk aluminate glasses is difficult due to the high melting temperatures and high crystallisation rates characteristic for this system, which requires ≈10^{2-3} K/s cooling rates to preserve the glassy state. Preparation of aluminate glasses, especially those with high Al_2O_3 content is therefore both technologically and economically demanding. Considering the above, in this work we chose the $62{\cdot}5Al_2O_3.xLa_2O_3.(37{\cdot}5-x)Y_2O_3$ glass system to study the mixed rare-earth effect on micro-hardness, viscosity and coefficient of thermal expansion by partial substitution of lanthanum for yttrium by laser-heated aerodynamic levitator (ADL) technique.

2. Experimental procedure

2.1. Sample preparation

High-purity Y_2O_3, La_2O_3 and Al_2O_3 powders (Aladdin, purity: >99·99%) were mixed stoichiometrically with the composition $62{\cdot}5Al_2O_3.xLa_2O_3.(37{\cdot}5-x)Y_2O_3$ in molar percent. The mixtures were firstly compacted into pellets with a diameter of 20 mm under a pressure of 150 MPa and then sintered at 1373 K for 5 h in an air atmosphere. After that, about 25 mg of crushed sintered pellet was levitated by O_2 gas and heated by 80 and 40 W CO_2 lasers in the conical nozzle of an ADL. To obtain a completely homogeneous melt, the sample is kept above the melting point for a few seconds. Subsequently, the droplet was cooled to room temperature at a rate of 500~1000 K/s by turning off the laser power. From the slope of the cooling curve, we can get the cooling rate of the sample (Figure 1(a)). The obtained glass with ~2 mm diameter is annealed for 2 h at their respective annealing points (=1012·3 Pa s). The chemical compositions of the final glasses were determined using x-ray fluorescence (Table 1). The glass nature of the sample is confirmed by using a x-ray diffractometer (Rigaku, RU-200B) with Cu K_α radiation (λ=1·5418 Å). The glass density ρ was determined using helium ideal-gas pycnometry, (ultrapyc 1200e, Quantachrome). The measurements were conducted five times for each sample.

Table 1. Analysed chemical composition of the glasses understudy. Compositions were analysed by x-ray fluorescence (±0·2 mol%)

Sample ID	Composition(mol%) Al_2O_3	La_2O_3	Y_2O_3
r=0·0	62·4	/	37·6
r=0·2	61·7	7·7	30·6
r=0·4	62·.1	15·1	22·8
r=0·6	61·9	22·5	15·6
r=0·8	62·3	30·6	7·1
r=1·0	62·2	37·8	/

2.2. Vickers and Knoop microhardness

Prior to indentation measurements, each glass sample was embeded by coagulated resin and polished carefully to obtain a mirror-like surface. Vickers hardness (H_V) and Knoop hardness (H_K) were determined using a micro-hardness instrument (Q10A+, QNESS in Austria) under ambient conditions (RH≈45%, T=28°C). Hardness is produced with a loading time of 10 s at an applied indentation load of 9·8 N, and a total of 20 indents were obtained on each sample. Then we measured the concave diagonal length of the indents without corner cracks immediately using an optical microscope. H_V is given by the following equation:

$$H_V=2P\sin\tfrac{1}{2}(138°)/d^2 \quad (1)$$

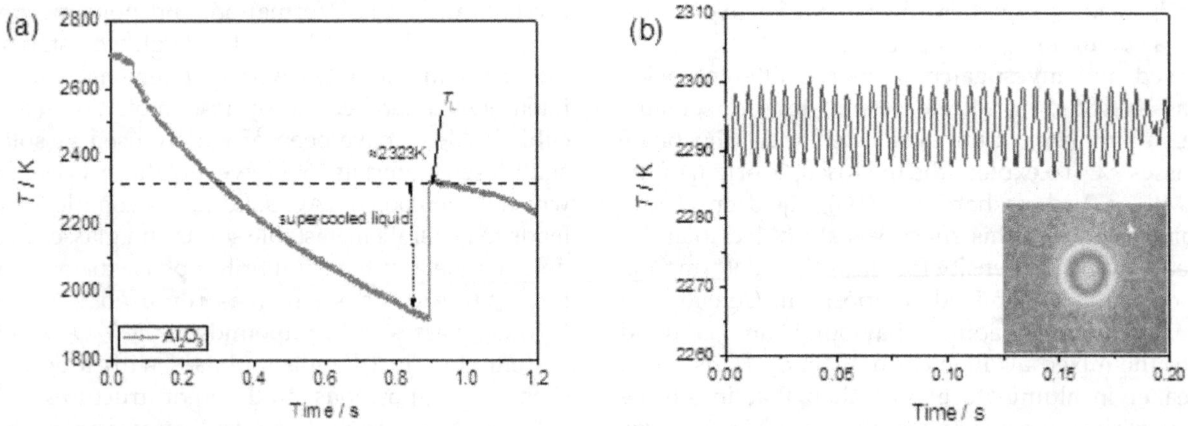

Figure 1. (a) Pyrometer traces for the Al_2O_3 samples showing supercooling and recalescence, when quenched rapidly at 2800 K by instantaneously cutting the laser power. (b) Temperature variation curve of Al_2O_2 at 2300 K, which is recorded by Pyrometer. Inset: thermal infrared image of Al_2O_3 droplet at 2300 K measured by MCS640 infrared thermography [Colour available online]

Where P is the indentation load (N) and d refers to the concave diagonal length of the indent (μm).

H_K was calculated according to the equation:

$$H_K=14{\cdot}229P/d^2 \quad (2)$$

where P and d refer to the indentation load and the long concave diagonal length, respectively. To evaluate the C_R to radial cracks, the number of indentation corners is divided by the number of cracks for each indentation, and the results are averaged based on the number of indentation tests. The curves of cracking probability are generated by plotting the cracking probability against the applied loading force. By examining these cracking probability curves, the load at which the probability reaches 50% is identified as C_R.[20,21]

2.3. Calorimetric analysis

Differential scanning calorimetry (DSC 449 F1, Netzsch) was used to determine glass transition temperature (T_g) and crystallisation temperature (T_c) in Ar_2 atmosphere (40 ml/min flow rate) at a heating rate of 10 K/min. The glass was pressed into a powder and held in a lidded Pt crucible at 313 K for 5 min, and upscanned at 10 K/min to 1500 K (see Figure 4(a)) to obtain T_c. Based on the first upscan result, a new sample was upscanned at 10 K/min to the temperature of the offset of glass transition peak (see Figure 4(b)) and then downscanned to 323 K at the same rate. The T_g was determined from the second upscan curve of the sample at the same heating rate of 10 K/min.

2.4. Raman spectroscopy

Raman spectra were recorded at room temperature on optically polished glass samples, which is embed by coagulated resin, surfaces by HORIBA LabRAM HR Evolution micro-Raman spectrometer. We collected spectra in the range of 100 to 1200 cm^{-1} under the conditions: 0·5 W 532 nm green laser power, 1 s integration time, 60 accumulations. We have removed the spectral background by a simple polynomial baseline, then normalised to the integrated signal intensity.

2.5 Thermophysical properties of high temperature liquids

The high-temperature thermophysical properties of the liquids were measured using the advanced ADL apparatus between ultra-high (up to 3000°C) and the supercooled liquid temperatures. Heating facilitated by 80 W and 40 W CO_2 SynRAD Evolution Series 10·59 μm lasers. A top and bottom configuration is chosen to enable increased temperature homogeneity. Temperature measurements are carried out using a high speed single-colour pyrometer at wavelengths up to 0·9 μm. As the emissivity of the sample is unknown, we assume the sample emissivity as a constant and calibrate the temperature measured using Al_2O_3 based on Wien's law:[22,23]

$$\frac{1}{T}-\frac{1}{T_P}=\frac{1}{T_L}-\frac{1}{T_{L,P}} \quad (3)$$

where T is the absolute temperature. T_P is the measured temperature by pyrometer. The T_L is the melting point of Al_2O_3 (2323 K).[22,23] $T_{L,P}$ corresponds to the maximum of recalescence as shown in Figure 1(a). After turning off the laser, the melt temperature drops below the melting point, and it increases rapidly to the melting point because of the release of the latent heat when crystallisation begins, which is called recalescence. The temperature does not change at the melting point as long as the melt coexists with crystals. The dependent time of molten Al_2O_3 at a given temperature is also shown in Figure 1(b) to estimate the stability of calibrated temperature. We also measured the thermal infrared image of Al_2O_3 droplet at 2300 K by MCS640 infrared thermography. The absolute temperature uncertainty was estimated as ±50 K.

To measure density of molten droplets, the important prerequisite is the temperature dependent of the sample shape. However, the droplet is usually partially visible since the bottom part hidden by the edge of the conical nozzle. Here, the ellipse droplet is assumed rotational symmetry around the vertical axis. The density ρ of the floating liquid drop can be calculated as,

$$\rho=3M/(4\pi r_m r_M^2) \quad (4)$$

where r_M and r_m are the half of major and minor axis of the ellipsoid, respectively. M is the weight of the ellipsoid. The precision milled steel ball bearings in different sizes and ruby spheres in 2 mm diameter are used to the camera calibration. Based on the density, the thermal expansion coefficient can be calculated as,

$$\beta=\frac{1}{V}\cdot\frac{dV}{dT}=-\frac{1}{\rho}\cdot\frac{d\rho}{dT} \quad (5)$$

where β is the volume expansion coefficient. The expansion coefficient for glasses is determined in one direction is the linear thermal expansion coefficient. The β=3* linear thermal expansion coefficient.

Based on the relationship between the resonance frequency, the decay constant and the shape size of the levitating drop, the viscosity of melt can be measured accurately. To obtain the viscosity (η) value of a spherical droplet, oscillations are analysed. Because η is related to both the resonance frequency ν_R of the oscillating droplet and the damping constant of the oscillation. It is calculated based on the Rayleigh equations:[24]

$$\eta = \frac{1}{(l-1)(2l+1)}\frac{3M\Gamma}{4\pi r} \quad (6)$$

where r is the radius of the oscillating sphere. l describes the oscillation mode, in details, the sample is excited at its (l=2, m=0) mode, whereby projected sample shape changes are not influenced by rotation. G is the damping constant and can be determined by fitting the damping curve of horizontal sample radius.

$$r_{h/v}(t)=A\exp[-\Gamma(t-t_1)]\sin[2\pi\nu_R(t-t_0)]+r_{av} \quad (7)$$

where A is the amplitude, and t_1 and t_0 refer to the onset and offset of time decay, respectively.[7,22,24] To determine the volatilisation during the measurement, we measured the mass of the La_2O_3–Al_2O_3 spheres before and after measurement. The mass loss of the 22·3 mg sample was found to be about ~0·4 mg at 2300 K for about 5 min.

3. Results and discussion

3.1. Hardness and crack resistance

Hardness is usually determined by the micro-indentation method with Vickers, Knoop, Rockwell and Brinell. It is difficult to accurately measure the hardness of brittle materials since many factors, e.g. indentation load, indentation time, radial cracks, play roles in the measuring process. Most of the indentation hardness decreases with increasing indentation load, and this is known as the indentation size effect (ISE).[25] Figure 2(a) shows both the H_V and H_K as a function of indentation load for the

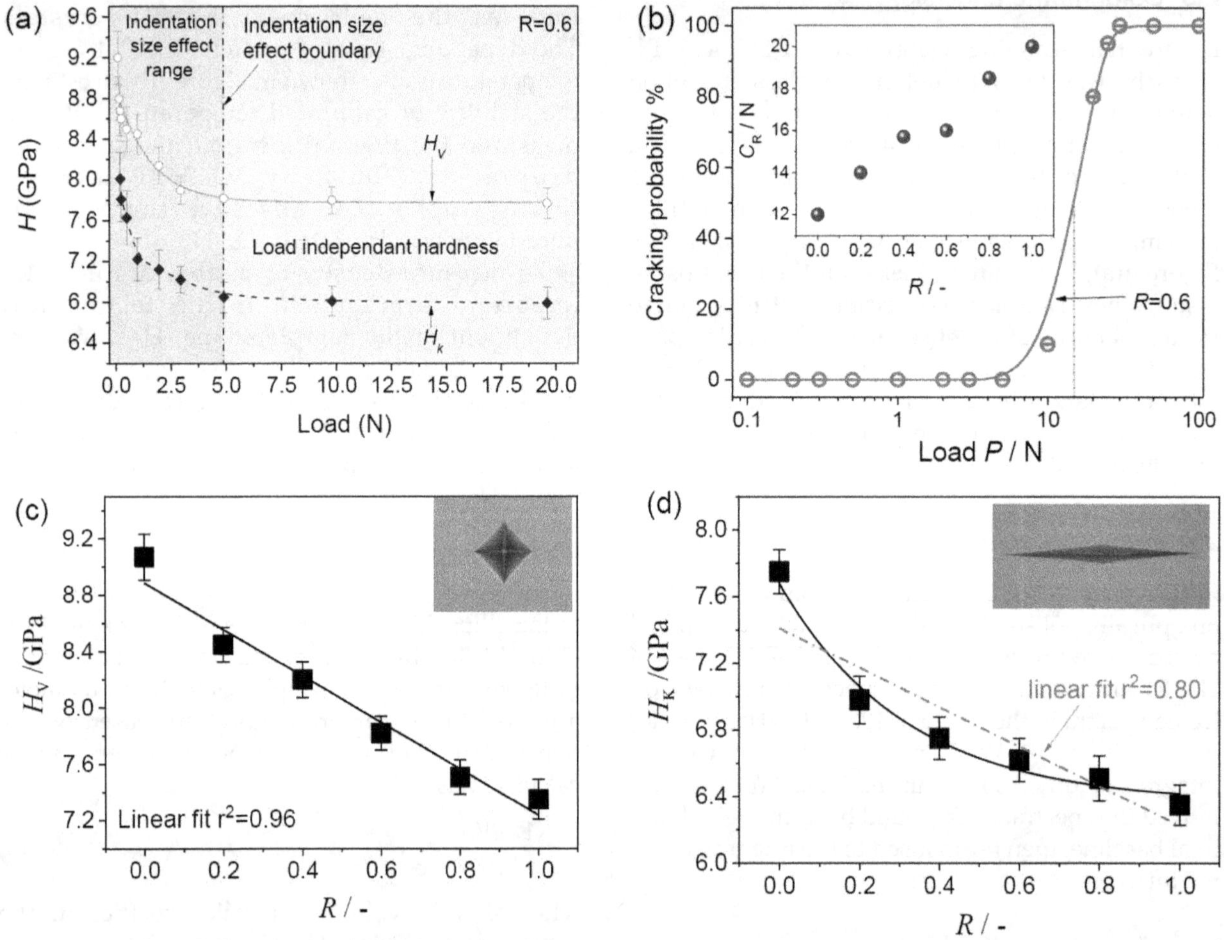

Figure 2. (a) H_V and H_K as a function of indentation load for R=0·6 (R=[La_2O_3]/([Y_2O_3]+[La_2O_3])) glass, showing an indentation size effect (ISE). Both lines are guides for the eye. (b) Dependence of cracking probability on the applied load P for the $Y_3Al_5O_{12}$ (R=0·6) glasses. Inset: Composition dependence of crack-resistant (C_R) as a function of the molar ratio R. Here C_R is the load at which the cracking probability corresponding to 50%. The solid line fitted by Weibull function represents the apparent relation between cracking probability and the load P. (c) H_V as a function of the molar ratio R in mixed rare-earth aluminate glasses. Solid line: linear relation between H and R (r^2=0·96). (d) H_K as a function of the molar ratio R in mixed rare-earth aluminate glasses. Solid line: Nonlinear relation between H and R, which is guide for the eye; whereas the dashed line is a linear fit to data (r^2=0·80). Inset: Optical images of Vickers indentation (c) obtained at 9·8 N and Knoop indent (d) obtained at 9·8 N on the surface of the as-prepared glasses. The error bar of the density is calculated by average deviation [Colour available online]

$62{\cdot}5Al_2O_3.xLa_2O_3.(37{\cdot}5-x)Y_2O_3$ glass with R=0·6 (R=[La_2O_3]/([Y_2O_3]+[La_2O_3])). Apparently, both H_V and H_K firstly decrease with the increasing applied loads to about 5 N. At sufficiently high indentation test loads of about 5 N, hardness reached to a plateau. Then, in the following micro-indentation experiments, the indentation load should be equal or greater than 5 N, which can avoid the indentation size effects.

In addition to the indentation load, cracking also affected the micro-indentation hardness. Here, we quantify the resistance to fracture by crack resistant (C_R), which is calculated through the crack probability curves fitted by Weibull's statistics (Figure 2(b)). C_R is a load corresponding to a crack probability of 50%, or the number of median cracks emanating from the four corners of the Vickers indent at impression.[10,26,27] As shown in the inset of Figure 2(b), the C_R exhibits an increase trend as Y_2O_3 is gradually substituted by La_2O_3. Based on the C_R results of rare earth aluminate glasses, the hardness values were determined at 9·8 N load to avoid cracking effect on hardness. The Vickers (H_V) microhardness values as a function of the molar ratio R were plotted in Figures 2(c) and (d), respectively.

For the rare-earth aluminate glasses, with the gradual substitution of Y_2O_3 by La_2O_3, H_V exhibits a linear decrease over the total composition range. Hardness is frequently related to the cationic field strength (CFS) of the modifier elements, whereby glasses incorporating the highest CFS ions have increased hardness. It should be noted that the variation of H_V with the composition in this work neither obey the mixed alkali, alkaline-earth effect nor the mixed modifier effect in the oxide glasses.[4,5,10,28] La^{3+} has a larger ionic radius (Shannon ionic radii for six-fold coordination 0·1061 nm) and smaller field strength than those of Y^{3+} (ionic radii for six-fold coordination 0·0892 nm). The CFS of the rare-earth element La^{3+} and Y^{3+} are 3·703 and 2·807 Å^{-2}, respectively.[29] It will result in the single bonding energy of Y–O bonds higher than that of La–O bonds in the rare-earth aluminate glasses. In addition, the packing density (C_g) were reported to increase with the decreasing ionic radius of rare-earth in aluminate glasses.[30,31] Also, the densities of the glasses at room temperature exhibit a linear decrease as a function of the molar ratio of [La_2O_3]/([Y_2O_3]+[La_2O_3]) in Figure 3.

The different roles played by the rare earth affect the vibrational properties in various ways in Raman spectra. Figure 7 clearly shows the intense bands of Raman scattering spectra in three regions: the lowest wavenumber region (200–550 cm^{-1}), the middle region (550–650 cm^{-1}) and the highest region (650–950 cm^{-1}). The strong band at 350 cm^{-1} could be assigned to transverse motions of the bridging Al–O

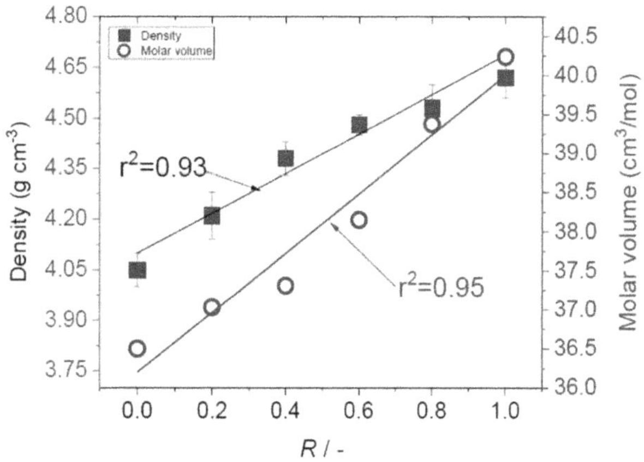

Figure 3. Density and molar volume as a function of the molar ratio [La_2O_3]/([Y_2O_3]+[La_2O_3]) for rare earth aluminate glasses. The solid line is the linear fitting of the measured data. The error bar of the density is calculated by average deviation. The errors are smaller than the size of the symbols for R=0·6. The inset show the photograph of the glasses [Colour available online]

linkages (corresponding the middle range structure of glass) and related to La–O bonding, while the high-frequency 780 cm^{-1} band may be assigned to aluminium–oxygen symmetric stretching vibrations of the AlO_4 tetrahedra groups.[13,32,33] The band in the middle region (580 cm^{-1}) are assigned to the transverse motion of bridged oxygen with the Al–O–Al linkage and related to AlO_5 or AlO_6. For the glasses, the lowest region bands shift to lower wavenumbers with a increase R. Raman peak width at half height of the lower-frequency region (denoted as PWLL) show two distinct compositional regimes with an increase R. Initially, as Y_2O_3 is substituted for La_2O_3, the PWLL increases, after which the PWLL decreases upon further substitution (Figure 4(b)). These results suggest that the disarray of middle range structure and La–O increases and then decrease with an increase in La_2O_3 content; In the middle region, a small peak appears in all spectra and the intensity decreases with an increase in La_2O_3 content. In the glasses, the peak of the highest region is located at 780 cm^{-1} and does not change significantly depending on the composition. However, Raman peak width at half height of the high-frequency region (denoted as PWHL) decreases with an increase R. Therefore, for the glass formation region, the local environment around AlO_2 does not change significantly depending on the amount of La_2O_3 of R=0·6, 0·8 and 1. However, the PWHL of the decreases with an increase in R, which is indicative of the Q_n (n is the bridging oxygen atoms number) orderly with an increase R (Figure 4(c)). In other word, mixed rare earth decrease the O–Al–O bond angle distribution.

The Raman results indicated that the densification of network increased with the decreasing ionic radii of the rare-earth cations

(Figure 4(a)). Raman spectroscopy high wavenumber vibrational modes may be associated with higher network connectivity because high wavenumber vibrational is required more energy to induce atomic vibrations. In this place, the lower frequency band is shift to higher wavenumbers results from an densification of the glass network (Figure 4(a)). Figure 4(d) shown a stronger correlation between the middle range structure of glasses and the C_R of aluminate glasses. Indeed, the result implies that the facilitate local densification of network can reduction of the residual stress around the deformation zone during indentation. Therefore, we infer that both the densification of network and C_g decrease as Y_2O_3 is substituted by La_2O_3 result in the decrease of H_V and increase of C_R.

3.2. Thermophysical properties of rare-earth aluminate melts

Figure 5(a) shows the DSC curves of the rare-earth aluminate glasses. The glass transition temperatures (T_g) of glasses were determined from the second DSC upscan curve (as seen in the inset of Figure 5(b)). T_c is defined as the crystallisation peak temperature as denoted in the inset of Figure 4(a). From the previous studies, T_g values regarding the mixed modifier effect in any glass systems all show negative deviations from linearity.[3,5,7,10,34] For the rare-earth aluminate series, T_g of the La-containing end-member is lower than that of the Y_2O_3–Al_2O_3 glass. As Figure 4(a) shows, the Raman spectra of the Y_2O_3–Al_2O_3 glass has a higher densification network in the medium-range structure than that of La_2O_3–Al_2O_3 glass, i.e. more rigid constraints of Y_2O_3–Al_2O_3 glass, leading to a higher T_g of Y_2O_3–Al_2O_3 glass than La_2O_3–Al_2O_3 glass. Except that, the local environment of aluminium in La_2O_3–Al_2O_3 and Y_2O_3–Al_2O_3 glasses is investigated by the MAS NMR spectroscopy. The results suggest that four-fold Al content is approximately 90% and 60% in La_2O_3–Al_2O_3 and Y_2O_3–Al_2O_3 glasses, respectively.[32] Then, the network polymerisation of La_2O_3–Al_2O_3 glass is higher than Y_2O_3–Al_2O_3 glass. Second, the Raman

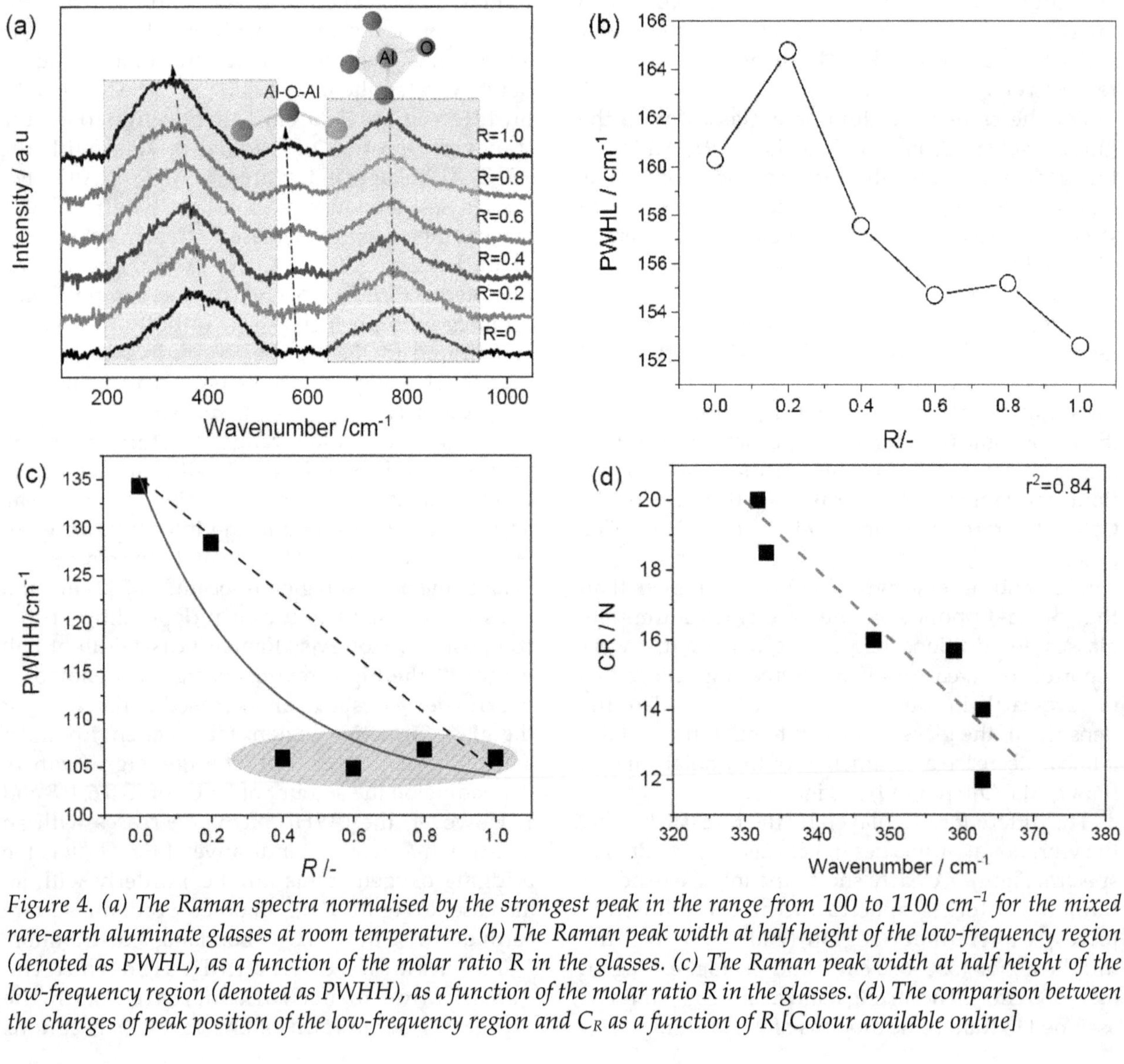

Figure 4. (a) The Raman spectra normalised by the strongest peak in the range from 100 to 1100 cm^{-1} for the mixed rare-earth aluminate glasses at room temperature. (b) The Raman peak width at half height of the low-frequency region (denoted as PWHL), as a function of the molar ratio R in the glasses. (c) The Raman peak width at half height of the low-frequency region (denoted as PWHH), as a function of the molar ratio R in the glasses. (d) The comparison between the changes of peak position of the low-frequency region and C_R as a function of R [Colour available online]

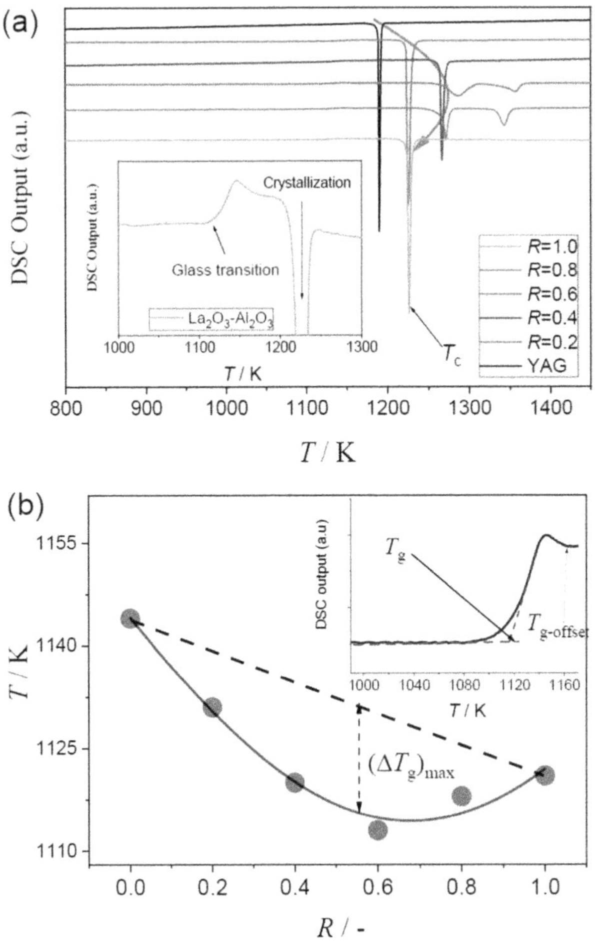

Figure 5. (a) The differential scanning calorimeter (DSC) output of the mixed rare-earth aluminate glass versus temperature (T). Inset: the DSC curve of the R=1 sample. (b) Composition dependence of the glass transition temperature T_g, obtained from DSC measurements by the standard method exampled as shown in the insert for aluminate glasses as a function of molar ratio R=[La_2O_3]/([Y_2O_3]+[La_2O_3]). Solid lines represent the nonlinear relation between T_g and R, whereas dashed lines are linear extrapolations between the end-member compositions; both are intended as guides for the eye. (ΔT_g) max is referred to the largest discrepancy between the linear and apparent relations, i.e. the largest mixed rare earth effect. The T_g, DSC uncertainty is about ±2–3 K [Colour available online]

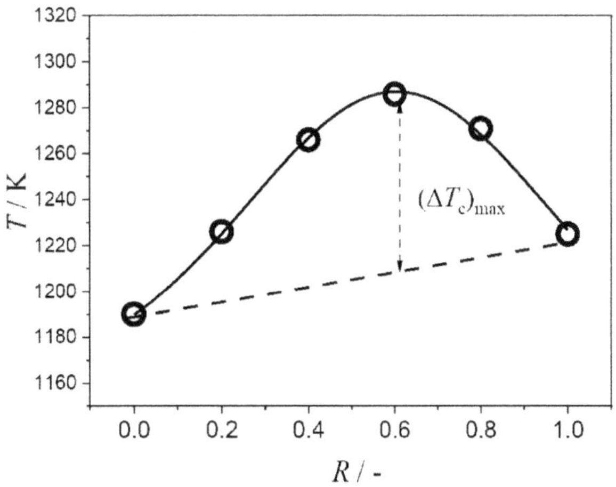

Figure 6. Composition dependence of the crystallisation temperature T_c, obtained from DSC measurements for aluminate glasses as a function of molar ratio R=[La_2O_3]/([Y_2O_3]+[La_2O_3]). Solid lines represent the nonlinear relation between T_c and R, whereas dashed lines are linear extrapolations between the end-member compositions; both are intended as guides for the eye. (ΔT_c) max is referred to the largest discrepancy between the linear and apparent relations, i.e. the largest mixed rare earth effect [Colour available online]

peak width at half height of the lower-frequency region (denoted as PWLL) show two distinct compositional regimes with an increase of R (Figure 4(b)). These results suggest that the disarray of middle range structure and La–O increases and then decrease with an increase in La_2O_3 content; Then, the T_g of glasses show a negative deviation from linearly could with the medium-range order structure densification network and bond energy.

The thermal stability is an important factor to be considered in developing high-quality aluminate optical glasses. In Figure 6, crystallisation temperature T_c is plotted as a function of the molar ratio of R [La_2O_3]/([La_2O_3]+[Y_2O_3]). We note that as Y_2O_3 is substituted for La_2O_3, the T_c increases. This trend continues until a substitution degree of 0·60 after which the T_c decreases upon further substitution. The largest deviation from linearity (the dashed line) is observed at a substitution degree of~0·60, i.e. the largest mixed rare earth effect is observed at a composition close to the one with equal concentrations of the two rare earth ions. The glasses were obtained from different cooling rate of the levitated melt when the laser power was turned off. The cooling curves for $Y_3Al_5O_{12}$ sample as a representative are shown in the Figure 7(a). The inset shown the XRD results of the YAG samples with different cooling rate. Based on the cooling curve and XRD results, we can get the cooling rate of YAG from a melt transform into glass. The thermal stability is an important factor to be considered in developing high-quality aluminate optical glasses. Generally, a higher T_c–T_g value corresponds to a stronger inhibition to nucleation and crystal growth processes, i.e. the more thermally stable.[35] The T_c–T_g value of the aluminate glasses as a function of R are shown in Figure 7(b). The cooling rate is determined by turning off the laser. A maximum T_c–T_g value at R=0·6 is observed, indicating that mixed rare-earth can improve glass thermally stable in the aluminate system.

Figure 8(a) shows the high temperature dependence on the density of the 62·5Al_2O_3.$x$$La_2O_3$.(37·5–$x$)$Y_2O_3$ liquids with various R values in the

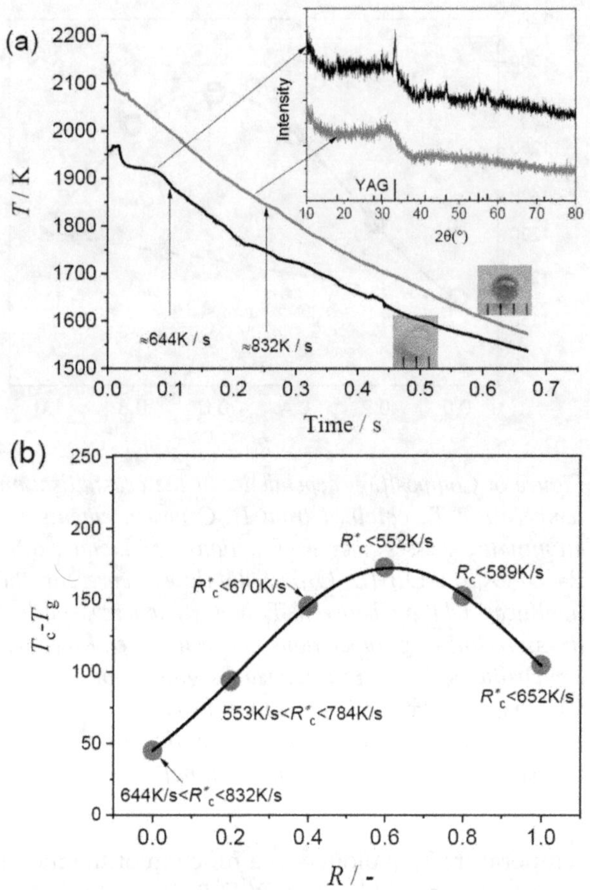

Figure 7. (a) The cooling curve for $Y_3Al_5O_{12}$ sample. Inset: XRD patterns for the $Y_3Al_5O_{12}$ glass with different cooling rates. (b) Composition dependence of the glass stability for aluminate glasses as a function of molar ratio R. R_c defined as the lowest cooling rate at which a melt transforms into glass are calculated by the cooling curve. The solid line represents the apparent relation between R_c* and the molar ratio of R, which are guides for the eye [Colour available online]*

range from 1800–2400 K. The density of the liquid exhibited a linear behaviour as a function of temperature and was linearly fitted, from which the average linear coefficients of thermal expansion (*CTE*) of the aluminate liquids were determined. The dependence of *CTE* on $[La_2O_3]/([La_2O_3]+[Y_2O_3])$ deviates positively from linearity (dashed line) (Figure 8(b)), exhibiting the mixed rare-earth effect. The *CTE* of a glass-forming liquid is contributed from both the configurational and vibrational components, which also directly connected with the network topology.[36] In most condition, a higher configurational entropy corresponds to a higher topological freedom degree, resulting in a higher *CTE*.[36–38] The configurational entropy (S_{conf}) is composed of topological and chemical contributions:[39,40] $S_{conf}=S_{topo}+S_{mix}$. In the aluminate glasses, S_{conf} arises from the disorganised distribution of tetrahedral species with various Al–O bond distances and Al–O–Al bond angles, and from the mixing of different cations in similar sites, respectively (Figure 4(b)), which consequently increase the configurational entropy. However, a random mixing of rare earth decrease the O–Al–O bond angle distribution (Figure 4(c)). Hence, the positive divergence from the linear of configurational entropy and angle around oxygen in the cation–oxygen–cation bond in the glass network could be the reason of liquid *CTE* exhibiting a nonlinear trend.

The viscosity (η) of the melt plays a key role in the manufacture of glass, especially the fining, moulding and annealing processes. Because η is related to the resonance frequency ν_R and mass *M*, the acoustically driven excitation sweeps for samples were conducted. An example of a fundamental resonance peak is clearly visible centred at 177 Hz. Where the oscillating horizontal

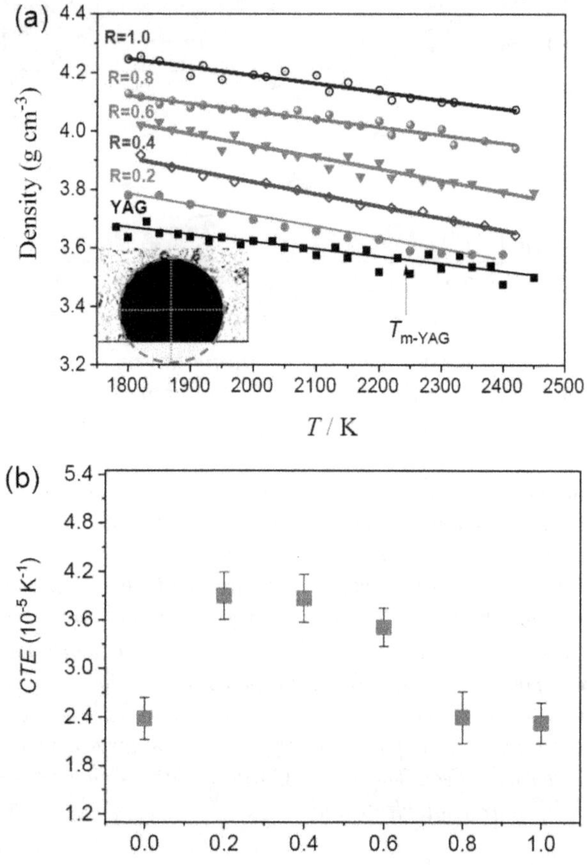

Figure 8. (a) Density of mixed rare earth aluminate melts as function of temperature. Inset: Size and shape of aerodynamically levitated droplet obtained by contrasting backlit imaging. The uncertainties for the present density measurements data are estimated to be about 5%. (b) Composition dependence of CTE as a function of the molar ratio $R=[La_2O_3]/([Y_2O_3]+[La_2O_3])$ in the temperature range from 1800–2400 K for the $62{\cdot}5Al_2O_3.xLa_2O_3.(37{\cdot}5-x)Y_2O_3$ melts. The error bar of the CTE considerate the underestimate sample size and fit error of density [Colour available online]

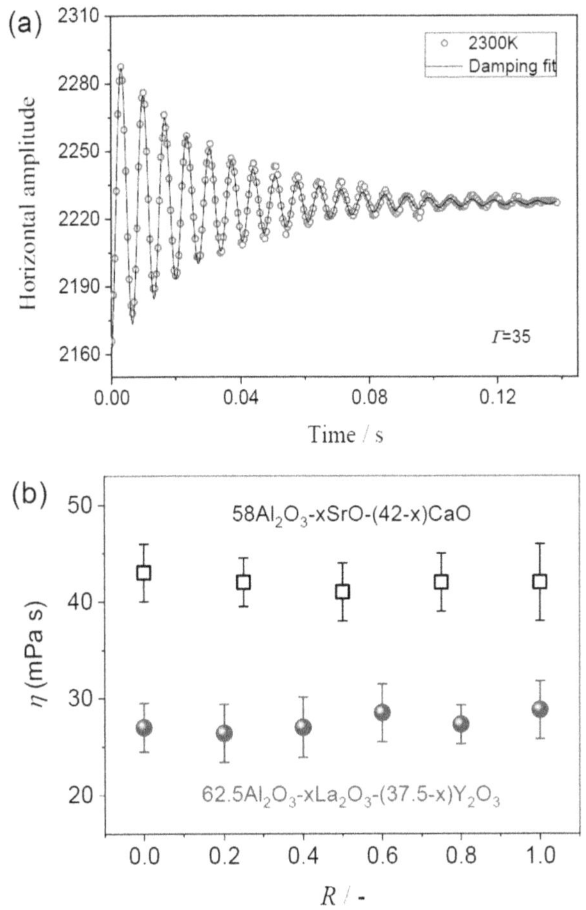

Figure 9. (a) Damping curve, together with the damped harmonic oscillator fit using Equation (4), according to the recorded radius data for the $Y_3Al_5O_{12}$ sample at 2300 K. (b) Composition dependence of the isothermal viscosity at T=2300 K for mixed rare-earth aluminate and mixed alkaline earth aluminate glasses[7] as a function of molar ratio R. The error bar of the viscosity are calculated by average deviation. The error bar of the viscosity is consideration of underestimate sample size, the error of damped harmonic oscillator fit and estimated to be about 20% of the measured value [Colour available online]

droplet amplitude obtained from image analysis together with the Fourier transform is given in Figure 8(a). As an example, the time-dependent horizontal diameter of liquid YAG droplet and fitted curve (fitted by Equation (7)) are shown in Figure 9(a). Figure 9(b) shows a near-linear trend of the viscosity determined by applying the oscillation droplet method at 2300 K with increasing R. With an increase in isokom temperature, i.e. a decrease in viscosity, the deviation from linearity becomes less pronounced and approaches zero (Figure 9(b)). In other words, the mixed rare-earth effect is declining as the temperature increases, which is consistent with the previous findings in alkaline earth aluminate glasses.[7] According to the flow mechanism, η is believed to be controlled by the exchange of Al–O bonds of aluminate liquids. In rare earth aluminate liquids, the viscous flow is related to the breaking and creating processes of La(Y)–O and Al–O bonds.[41] In addition, the viscosity is also correlated with the ionic diffusion characteristics according to the Stokes–Einstein equation, especially at the temperature range higher than $1{\cdot}2T_g$ for a glass forming melt.[42] The bond energy and ionic diffusion activation energy for La–O band and Y–O bond is so weak at high temperature that the cationic potential energy mismatch effect is quietly faint. Hence, this could be the reason that the η of mixed rare-earth liquids have the near-linear trend at 2300 K, but a negative deviation from linear at T_g.

4. Conclusion

We studied the mixed rare-earth effect in a series of aluminate glasses by substituting Y_2O_3 for La_2O_3, using a contactless laser-heated aerodynamic levitator. Through the systematic investigation of Vickers and Knoop microhardness, we observed a near-linear trend in Vickers hardness similar with previous reported mixed alkaline-earth aluminate glasses, but different with other mixed alkaline-earth glasses. However, the Knoop microhardness show a near-linear trend in hardness with composition. The mixed rare-earth effects manifests itself as a maximum in glass stability and *CET* at high temperatures connect with the Al–O and La,Y–O linker not with the O–Al–O bond angle distribution. Furthermore, we demonstrate that the high crack resistance of aluminate glass is a stronger correlation with the medium-range structure Al–O linker by Raman spectra. The T_g show a nonlinear composition dependence, but the viscosity determined by applying the oscillation droplet method at 2300 K with increasing R shows a near-linear trend. Hence, it is obviously that the mixed rare-earth effect is declining as the temperature increases.

Acknowledgments

This work is financially supported by the National Natural Science Foundation of China (Nos: 52302003 and 52202006), National Natural Science Foundation of China（Nos: 52372006 and 52472005）Taishan Youth Scholar Project of Shandong Province (tsqn202103098), and the Colleges and Universities New Twenty Terms Foundation of Jinan City (No. 202333073). We are grateful for valuable discussions with Prof. Yuanzheng Yue (Aalborg University).

Authorship contribution statement

Zhitao Shan: data curation, preparation, writing-original draft; Yanfei Zhang and Qiuju Zheng: validation, writing-original draft; Shujiang Liu and Xiujun Han: visualization, supervision investigation, writing-reviewing and editing; Haizheng Tao: writing-reviewing and editing, funding acquisition validation; and Jiayan Zhang: supervision, writing-reviewing and editing.

Competing interests

The authors declare that they have no known competing financial interests or personal relationships that could have appeared to influence the work reported in this paper.

Data and code availability

All data needed to evaluate the conclusions in the paper are present in the paper.

References

1. Kurkjian, C. R. & William, R. P. Perspectives on the history of glass composition. *J. Am. Ceram. Soc.*, 1998, **81**, 795–813.
2. Yue, Y., Wang, M., Smedskjaer, M. M., Mauro, J. C., Sant, G. & Bauchy, M. Thermometer effect: origin of the mixed alkali effect in glass relaxation. *Phys. Rev. Lett.*, 2017, **119**, 095501.
3. Kjeldsen, J., Smedskjaer, M. M., Mauro, J. C., Youngman, R. E., Huang, L. & Yue, Y. Mixed alkaline earth effect in sodium aluminosilicate glasses. *J. Non-Cryst. Solids*, 2013, **369**, 61–8.
4. Shan, Z., Qiao, A., Liu, S. & Tao, H. Mixed intermediate effect on mechanical and rheological performances in Zn–Mg silicate glasses. *J. Alloy. Compound.*, 2018, **747**, 738–46.
5. Wilkinson, C. J., Potter, A. R., Welch, R. S., Bragatto, C., Zheng, Q., Bauchy, M., Affatigato, M., Feller, S. A. & Mauro, J. C. Topological origins of the mixed alkali effect in glass. *J. Phys. Chem. B*, 2019, **123**, 7482.
6. Terny, S. & Frechero, M. A. Understanding how the mixed alkaline-earth effect tunes transition metal oxides-tellurite glasses properties. *Phys. B*, 2020, **583**, 412054.
7. Shan, Z., Liu, S., Tao, H. & Yue, Y. Mixed alkaline earth effects on several mechanical and thermophysical properties of aluminate glasses and melts. *J. Am. Ceram. Soc.*, 2019, **102**, 1128–36.
8. Sato, R. K., Kirkpatrick, R. J. & Brow, R. K. Structure of Li, Na metaphosphate glasses by ^{31}P and ^{23}Na MAS-NMR correlated with the mixed alkali effect. *J. Non-Cryst. Solids*, 1992, **143**, 257–64.
9. Kamitsos, E. I. & Yiannopoulos, Y. D. Far-infrared spectra of alkali germanate glasses and correlation with electrical conductivity. *Phys. Rev. B*, 1996, **54**, 9777.
10. Shan, Z., Zhang, Y., Liu, S., Tao, H. & Yue, Y. Mixed-alkali effect on hardness and indentation-loading behavior of a borate glass system. *J. Non-Cryst. Solids*, 2020, **548**, 120314.
11. Kjeldsen, J., Smedskjaer, M. M., Mauro, J. C. & Yue, Y. Hardness and incipient plasticity in silicate glasses: Origin of the mixed modifier effect. *App. Phys. Lett.*, 2014, **104**, 051913.
12. Hervé, L., Rouxel, T., Fargeot, D., Besson, J. L. & Piriou, B. Yttrium SiAlON glasses: structure and mechanical properties -elasticity and viscosity. *J. Non-Cryst. Solids*, 1996, 201, 128–45.
13. Kumar, M. V., Ishikawa, T., Basavalingu, B., Okada, J. T. & Watanabe. Y., Thermal and optical properties of glass and crystalline phases formed in the binary R_2O_3–Al_2O_3 (R=La–Lu and Y) system. *J. App. Phys.*, 2013, **113,** 19.
14. Salaheddine, A., Deschamps, M., Lambert, J., Suchomel, M. R., Meneses, D. D. S., Matzen, G., Massiot, D., Véron, E. & Allix, M. Crystallisation of Y_2O_3–Al_2O_3 rich glasses: synthesis of YAG glass-ceramics. *J. Phys. Chem. C*, 2011, **115**, 20499–506.
15. Ma, X., Li, X., Li, J., Genevois, C., Ma, B., Etienne, A., Wan, C., Véron, E., Peng, Z. & Allix, M. Pressureless glass crystallization of transparent yttrium aluminum garnet-based nanoceramics. *Nature Commun.*, 2018, **9**, 1175.
16. Aasland, S. & McMillan, P. F. Density driven liquid–liquid phase separation in the system Al_2O_3–Y_2O_3. *Nature*, 1994, **369**, 633–6.
17. Koyama, C., Ishikawa, T. & Watanabe, Y. Density measurements of molten $Y_3Al_5O_{12}$ and Y_2O_3–Al_2O_3 compounds using an electrostatic levitation technique. *J. Am. Ceram. Soc.*, 2023, **106** (10), 5648–53.
18. Skinner, L. B., Barnes, A. C., Salmon, P. S., Crichton, W. A. Phase separation, crystallisation and polyamorphism in the Yl_2O_3–All_2O_3 system. *J. Phys.: Condens. Mater.*, 2008, **20**, 205103.
19. Watanabe, Y., Masuno, A., Inoue, H. Glass formation of rare earth aluminates by containerless processing. *J. Non-Cryst. Solids*, 2012, **358**, 3563–6.
20. Rosales-Sosa, G. A., Masuno, A., Higo, Y. & Inoue, H. Crack-resistant Al_2O_3–SiO_2 glasses. *Sci. Rep.*, 2016, **6**, 1–7.
21. Shan, Z. T., Zhao, T. Y., Ke, X. F., Ren, J. J., Tao, H. Z. & Yue, Y. Z. Oxygen tri-clusters make glass highly crack-resistant. *Acta. Mater.*, 2024, **262**, 119425.
22. Langstaff, D., Gunn, M., Greaves, G. N., Marsing, A., Kargl, F. Aerodynamic levitator furnace for measuring thermophysical properties of refractory liquids. *Rev. Sci. Instrum.*, 2013, **84**, 124901.
23. Rayleigh. L. On the capillary phenomena of jets. *Proc. R. Soc. London*, 1879, **29**, 71–97
24. Liu, , H., Chen, W., Pan, R., Shan, Z., Qiao, A., Drewitt, J. W. E., Hennet, l., Jahn, S., Langstaff, D. P. & Chass. G. A. From molten calcium aluminates through phase transitions to cement *Phases*. *Adv. Sci.*, 2019, **7**, 1902209.
25. Bernhardt, E. O. Uber die Mikrohärte der Feststoffe im Grenzbereich des kickschen hnlichkeitssatzes. *Zeit. für Metallkunde*, 1941, **33**, 135–9.
26. Ke, X., Shan, Z., Li, Z., Tao, Y., Yue, Y. & Tao. H. Toward hard and highly crack resistant magnesium aluminosilicate glasses and transparent glass-ceramics. *J. Am. Ceram. Soc.*, 2020, **103**, 3600–9.
27. Bradtmuller, H., Uesbeck, T., Eckert, H., Murata, T., Nakane, S. & Yamazaki. H. Structural origins of crack resistance on magnesium aluminoborosilicate glasses studied by solid-state NMR. *J. Phys. Chem. C*, 2019, **123** (24), 14941–54.
28. Shan, Z. T., Li, C. J., Tao, H. Z. Mixed alkaline earth effect on the mechanical and rheological properties of Ca–Mg silicate glasses. *J. Am. Ceram. Soc.*, 2017, **100**, 4570–80.
29. Lv, P., Wang, C., Stevensson, B., Yue, Y., Wang, T. & Edén, M. Impact of the cation field strength on physical properties and structures of alkali and alkaline-earth borosilicate glasses. *Ceram. Int.*, 2022, **48**, 18094–107.
30. Shelby, J. E. & Kohli. J. T. Rare-earth aluminosilicate glasses. *J. Am. Chem. Soc.*, 1990, **73**, 39–42.
31. Sosa, R., Gustavo, A., Masuno, A., Higo, Y., Watanabe, Y. & Inoue, H. Effect of rare-earth ion size on elasticity and crack initiation in rare-earth aluminate glasses. *J. Am. Ceram. Soc.*, 2018, **101**, 5030–6.
32. Watanabe, Y., Masuno, A., Inoue, H., Yanaba, Y. & Katoa, K. Influence of modifier cations on the local environment of aluminum in La_2O_3–Al_2O_3 and Y_2O_3–Al_2O_3 binary glasses. *Phys. Chem. Chem. Phys.*, 2020, **22** (35), 19592–9.
33. Jaworski, A., Stevensson, B. & Eden, M. The bearings from rare-earth (RE=La, Lu, Sc, Y) cations on the oxygen environments in aluminosilicate glasses: a study by solid state ^{17}O NMR, molecular dynamics simulations, and DFT calculations. *J. Phys. Chem. C*, 2016, **120**, 13181–98.
34. Shan, Z., Wang, H., Wang, Y., Shen, Y., Zhang, J., Chen, Y., Zheng, Q., Zhang, Y. & Tao, H. Mixed alkaline earth effects on crystallization behavior of basalt glasses and liquids. *J. Alloys Compound*, 2021, **874**, 159986.
35. Zheng, Q., Mauro, J. C., Smedskjaer, M. M., Youngman, R. E., Potuzak, M. & Yue, Y. Glass-forming ability of soda lime borate liquids. *J. Non Cryst. Solids*, 2012, **358**, 658–65.
36. Marcel, P., Mauro, J. C., Kiczenski, T. J., Ellison, A. J. & Allan. D. C. Communication: resolving the vibrational and configurational contributions to thermal expansion in isobaric glass-forming systems. *J. Chem. Phys.*, 2010, **133**, 91102.
37. Senapati, U. & Varshneya, A. K. Configurational arrangements in chalcogenide glasses: a new perspective on Phillips' constraint theory. *J. Non-Cryst. Solids*, 1995, **185**, 289–96.
38. Reis, R. M. C. V., Mauro, J. C., Geisinger, K. L., Potuzak, M., Smedskjær, M. M., Guo, X. & Allan, D. C. Relationship between viscous dynamics and the configurational thermal expansion coefficient of glass-forming liquids. *J. Non-Cryst. Solids*, 2012, **358**, 648–51.
39. Losq, C. L. & Neuville, D. R. Molecular structure, configurational entropy and viscosity of silicate melts: Link through the Adam and Gibbs theory of viscous flow. *J. Non-Cryst. Solids*, 2017, **463**, 175–88.
40. Neuville, D. R. & Richet, P. Viscosity and (Ca, Mg) mixing in molten pyroxenes and garnets. *Geochim. Cosmochim. Acta*, 1991, **55** (4), 1011–21.
41. Smedskjaer, M. M., Mauro, J. C., Kjeldsen, J. & Yue, Y. Microscopic origins of compositional trends in aluminosilicate glass properties. *J. Am. Ceram. Soc.*, 2013, **96**, 1436–43
42. Yusuke, M., Hidemori, T., Akai, N., Kawai, A., Shibuya, K., Koguchi, S. & Kitazume. T. EPR study of rotational diffusion in viscous ionic liquids: analysis by a fractional Stokes–Einstein–Debye law. *Chem. Lett.*, 2009, **38**, 124–5.

CONFERENCE DIARY

9–10 July
The Advanced Materials Show, Birmingham, UK.
Danny Scott, Event Director, Event Partners, Crown House, Crown Road, Portslade BN41 1SH, UK. Email danny.scott@event-partners.com Web advancedmaterialsshow.com

13–17 July
Optica Advanced Photonics Congress, including Specialty Optical Fibers, Marseille, France.
Optica Customer Service. Email custserv@optica.org Web www.optica.org

28–31 July
Sustainable Energy and Environmental Protection Conference. Including on 31 July the fifth symposium on decarbonisation and resource efficiency in the oxide materials sector, London, UK.
Dr Marialuisa Acquaviva, SEEP Conference, Brunel University, Uxbridge Campus, Kingston Ln, Uxbridge UB8 3PH, UK. Email seep@seepconference.com Web seepconference.com

10–14 August
17th International Conference on the Physics of Non-Crystalline Solids, Tsukuba, Japan.
Shinji Kohara, Measurement and Characterization, National Institute for Materials Science, 1-2-1 Sengen, Tsukuba, Ibaraki 305-0047, Japan. Email KOHARA.Shinji@nims.go.jp

24–28 August
EOSAM 2025 (European Optical Society Annual Meeting), Delft, The Netherlands.
EOSAM 2025 Email eosam@europeanoptics.org Web www.europeanoptics.org

31 August–4 September
XIXth Conference of the European Ceramic Society, Dresden, Germany.
Fraunhofer Institute for Ceramic Technologies and Systems IKTS, Winterbergstrasse 28, 01277 Dresden, Germany. Email info@zv.fraunhofer.de Web www.ecers2025.org

1–3 September
SGT Annual Conference, Cambridge, UK.
Christine Brown, Society of Glass Technology, 9 Churchill Way, Chapeltown, Sheffield S35 2PY, UK. Email christine@sgt.org Web www.sgt.org

1–3 September
SEMC 2025: the Ninth Int. Conf. on Structural Engineering, Mechanics and Computation (including: Structural Applications of Glass), Cape Town, South Africa.
Prof. Jan Belis, Ghent University, BelgiumEmail jan.belis@ugent.be Web semc.uct.ac.za

7–11 September
34th Annual Conference of the European Society for Biomaterials, Turin, Italy.
Organizing Secretariat, Via Roma 10, 16121 Genova, Italy. Email esb2025@aristea.com Web esb2025.org

8–12 September
23rd Congress of the AIHV, Mainz, Germany.
Email aihvsecretary@gmail.com Web aihv.org/

10–12 September
Glasspex/Glasspro India 2025, Mumbai, India.
Messe Düsseldorf India, 20th Floor, Tower A, Building No. 5, DLF Cyber City, Phase II, Gurugram - 122002, India. Email info@md-india.com Web glassproindia.com

11 September
AI and Energy Conservation – The Future of Sustainability, the 15th AIGMF conference, Mumbai, India.
Email info@aigmf.com Web aigmf.com/future-events.php#11

13–21 September
Venice Glass Week, Venice, Italy.
Email info@theveniceglassweek.com Web www.theveniceglassweek.com

15–19 September
Sumglass 2025, Wuhan, China.
Dr Lifeng Li, Conference Secretary. Email lifeng.li@whut.edu.cn Web sumglass2025.scimeeting.cn/en/web/index/25180

16–16 September
44th Annual Cement & Concrete Conference, Leeds, UK.
The Institute of Materials, Minerals & Mining on behalf of the Cementitious Materials Subgroup, 297 Euston Road, London NW1 3AD, UK. Web www.iom3.org

16–19 September
Vitrum 2025, Milan, Italy.
Email vitrum@vitrum-milano.it Web www.vitrumlife.it

28 September–1 October
MS&T25, Materials Science and Technology exhibition and conference incorporating ACerS 127th Annual Meeting, Columbus, Ohio, USA.
The American Ceramic Society, 550 Polaris Pkwy, Ste 510, Westerville, OH 43082, USA. Email customerservice@ceramics.org Web www.matscitech.org

3–4 October
ICON Stained Glass Group Annual Conference 2025, Canterbury, UK.
The Institute of Conservation, 106-109 Saffron Hill, London EC1N 8QS, UK. Web www.icon.org.uk

6–9 October
86th Glass Problems Conference and Symposium, Toledo, Ohio, USA.
Glass Manufacturing Industry Council, 550 Polaris Parkway, Suite 510, Westerville, OH 43082-7045, USA. Web gmic.org/glass-problems-conference/

19–23 October
Advanced Solid State Lasers Conference, Prague, Czech Republic
Optica Customer Service. Email custserv@optica.org Web www.optica.org

27–30 October
UNITECR 2025, Unified International Technical Conference on Refractories, Cancun, Mexico.
Email contact@unitecr2025.com Web unitecr2025.com

4–6 November
GlassBuild America, Orlando, Florida, USA.
National Glass Association, USA. Email attend@glass.org Web www.glass.org

6–9 November
Glasstech Asia and Fenestration Asia 2025, Jakarta, Indonesia.
Glasstech Asia Team. Email bauasia@mmiasia.com.sg and annie@mmiasia.com.sg Web glasstechasia.com.sg

10–12 November
M3P3: Materials, Minerals & Mining for People, Places, Planet, Birmingham, UK.

IoM3, 297 Euston Road, London NW1 3AD, UK. Web www.iom3.org

17–20 November
11th Intl. Symp. on Sustainable Materials Recycling Processes & Products, Lapu-Lapu, Philippines.
Web www.flogen.org

26–28 November
UZ Glass, Tashkent, Uzbekistan.
Email info@uzglass.com Web uzglass.com/

2026

17–22 January
SPIE Photonics West, San Francisco, California, USA.
Email customerservice@spie.org Web spie.org

12–16 April
SPIE Photonics Europe, Strasbourg, France.
Email customerservice@spie.org Web spie.org

12–16 April
ACerS Spring Meeting (including the Glass and Optical Materials Division Annual Meeting), Bellevue, WA, United States.
The American Ceramic Society, 550 Polaris Parkway, Suite 510, Westerville, OH 43082, USA. Email customerservice@ceramics.org Web www.ceramics.org

13–17 April
ICG Annual meeting/18th ESG Conference, Lyon, France.
USTV, 114 Rue la Boétie, 75008 Paris, France. Email ustv@ustverre.fr Web www.ustverre.fr

25–30 April
69th Annual SVC Technical Conference, Long Beach, California, USA.
Society of Vacuum Coaters (SVC), PO Box 10628, Albuquerque, NM 87184, USA. Email svcinfo@svc.org Web www.svc.org

20–21 May
Glassman Latin America, Mexico City, Mexico.
Ken Clark, Glassman Events. Email kenclark@quartzltd.com Web www.glassmanevents.com

6–10 July
11th International Conference on Borate Glasses, Crystals and Melts and Third International Conference on Phosphate Materials, Lille, France.
Gérald Lelong, Sorbonne Université - IMPMC, France. Web ustverre.fr/agenda/borate-xi-phosphate-iv/

27–31 August
International Festival of Glass 2026, St Helens, UK.
Glass Art Society, 700 NW 42nd St #101, Seattle, WA 98107, USA. Email info@glassart.org Web www.glassart.org

31 August–3 September
Sintering 2026, the International Conference on Sintering, Aachen, Germany.
Olivier Guillon, Forschungszentrum Jülich/RWTH Aachen University, Germany. Email sintering2026@fz-juelich.de Web www.sintering2026.org

6–11 September
ICC11, 11th International Congress on Ceramics, Sapporo, Japan.
ICC11 Office, The Ceramic Society of Japan, 2-22-17 Hyakunincho, Shinjuku, Tokyo 169-0073, Japan. Email ICC11@ceramic.or.jp Web www.ceramic.or.jp/en/icc11.html

17–22 September
ICOOPMA 2026. 11th International Conference on Optical, Optoelectronic and Photonic Materials and Applications, Hamamatsu, Japan.
Web wwp.shizuoka.ac.jp/icoopma2026/

23–25 September
GlassBuild America, Las Vegas, Navada, USA.

National Glass Association NGA, 8200 Greensboro Drive, Suite 302, VA 22102 McLean, USA. Email attend@glass.org Web www.glass.org

4–7 October
ACerS 128th Annual Meeting with Materials Science and Technology, Pittsburgh, PA, USA.
The American Ceramic Society, 550 Polaris Parkway, Suite 510, Westerville, OH 43082, USA. Email customerservice@ceramics.org Web www.ceramics.org

20–23 October
Glasstec 2026, Düsseldorf, Germany.
Anke Seltmann, glasstec, D-40474 Düsseldorf, Am Staad (Stockumer Höfe), Germany. Email seltmanna@messe-duesseldorf.de Web www.glasstec-online.com/

2027
1–6 May
70th Annual SVC Technical Conference, Dallas, Texas, USA.
Society of Vacuum Coaters (SVC), PO Box 10628, Albuquerque, NM 87184, USA. Email svcinfo@svc.org Web www.svc.org

23–27 July
16th International Conference on the Structure on Non-Crystalline Materials, Boulder, Colorado, USA.
Organisers: Jincheng Du (U North Texas), Sabyasachi Sen (UC Davis)

24–27 October
ACerS 129th Annual Meeting with Materials Science and Technology, Pittsburgh, PA, USA.
The American Ceramic Society, 550 Polaris Parkway, Suite 510, Westerville, OH 43082, USA. Email customerservice@ceramics.org Web www.ceramics.org

14th International Symposium on Crystallization in Glasses and Liquids, Guangzhou, China.
Organiser: Shifeng Zhou, South China University of Technology, Guangzhou, China.

Ninth International Workshop on Flow and Fracture of Advanced Glasses, France.

2028
6–11 May
71st Annual SVC Technical Conference, Nashville, Tennessee, USA.
Society of Vacuum Coaters (SVC), PO Box 10628, Albuquerque, NM 87184, USA. Email svcinfo@svc.org Web www.svc.org

23–27 July
ICC12, Int. Congress on Ceramics, Birmingham, UK.
Professor Jon Binner, School of Metallurgy and Materials, University of Birmingham, Edgbaston, Birmingham B15 2TT, UK. Email j.binner@bham.ac.uk

24–29 September
28th International Congress on Glass, Fukuoka, Japan.

22–25 October
ACerS 130th Annual Meeting with Materials Science and Technology, Pittsburgh, PA, USA.
The American Ceramic Society, 550 Polaris Parkway, Suite 510, Westerville, OH 43082, USA. Email customerservice@ceramics.org Web www.ceramics.org

2029
8–13 April
18th Pacific Rim Conference on Ceramic and Glass Technology including ACerS Annual Spring Meeting 2029, Waikoloa Village, HI, USA.
The American Ceramic Society, 550 Polaris Parkway, Suite 510, Westerville, OH 43082, USA. Email customerservice@ceramics.org Web www.ceramics.org

Arun and Darshana Varshneya endow University Conference on Glass

The Varshneya Endowed Glass Fund, under the stewardship of Alfred University, will support the University Conference on Glass, a biennial conference which is hosted on a rotational basis by four institutions: Alfred University, Missouri University of Science and Technology, Rensselaer Polytechnic Institute, and Pennsylvania State University. Funding will support the conference at all four host institutions. The conference series, established at Alfred University in 1970, will now be titled the Varshneya-University Conference on Glass.

"My global reputation originated from two sources: professorship at the New York State College of Ceramics at Alfred University and ownership of Saxon Glass Technologies, Inc.," Professor Varshneya said. "The latter itself came about from my faculty position. At a breakfast meeting with retired faculty, I had remarked to President Zupan that I had the greatest time at Alfred. In Frank Sinatra's words, 'Regrets I had a few, but too few to remember.' It was, therefore, natural for me to think about giving back."

Alfred University hosted the 21st annual University Conference on Glass in 2024. Undergraduate and graduate students, faculty, and industry professionals took part in the conference, learning about the latest developments in glass science, technology, and education. Sessions included introductory lectures by glass experts on important glass subjects and presentations by undergraduate and graduate students of their latest research.

In addition to supporting the Varshneya-University Conference on Glass, the Varshneyas' gift will also fund the Varshneya Lecture on Glass, an annual lecture series held at Alfred University. Varshneya said he expects the lecture series to bring renowned members of the glass science community to campus, similarly to the Samuel L. Scholes Memorial Lectures series, held annually at Alfred University in April. Established in 1982, the Scholes Lecture series honours the late Samuel R. Scholes, who in 1932 established the first glass science program in the United States at the New York State College of Ceramics at Alfred University.

"The subject matter (of the Varshneya Lecture on Glass) should include all glass science, technology, and education topics, building upon the fundamentals described in my textbook," he said, referring to Fundamentals of Inorganic Glasses, which he wrote at the request of his students. Now used as a textbook to teach glass science and technology to students worldwide.

Dr Arun Varshneya is professor of glass science and engineering, emeritus, whose teaching and research career at the New York State College of Ceramics at Alfred University spanned from 1982 to 2010. During his tenure, he taught nearly every course in glass science and engineering at the undergraduate and graduate levels. He is a Distinguished Life Member of the American Ceramic Society, an Honorary Fellow of the Society of Glass Technology (and Immediate Past-President) and a President's Award recipient for lifetime achievement by the International Commission on Glass.

Low temperature additive manufacturing of glass

Researchers have used a low temperature additive manufacturing process to build the glass cups. The optical behaviour of the printed cups can be tailored by altering the chemical components of the inks

In the same way as the sol-gel process opened up low temperature routes to glasses, additive manufacturing (3D printing) holds promise for fabricating complex glass structures that would be unattainable by traditional glass manufacturing routes. **Lincoln Laboratory**'s approach allows the additive manufacturing of multimaterial glass items without the need for high temperature processing. This low temperature technique could facilitate the widespread adoption of 3D printing for glass devices such as microfluidic systems, free-form optical lenses or fibres, and high temperature electronic components.

Additive manufacturing has expanded the ability to rapidly and cost-effectively create system components. However, traditional materials, various plastic and metal composites, can produce 3D printed structures with mechanical, chemical and thermal instabilities. Using inorganic composite glasses solves many of the instability issues and offers a promising approach to create structures with diverse shapes and properties. Via direct ink writing, Lincoln Laboratory uses custom inks from a variety of inorganic materials combined with a silicate solution to engineer the optical, chemical and electrical properties of a composite. The ink ingredients are widely available and direct ink writing allows the generous geometric freedoms to create glass structures. The curing process to generate a stable glass structure requires only 250°C instead of typical glass processing of >1000°C. Following curing in a mineral oil bath the components then just need washing.

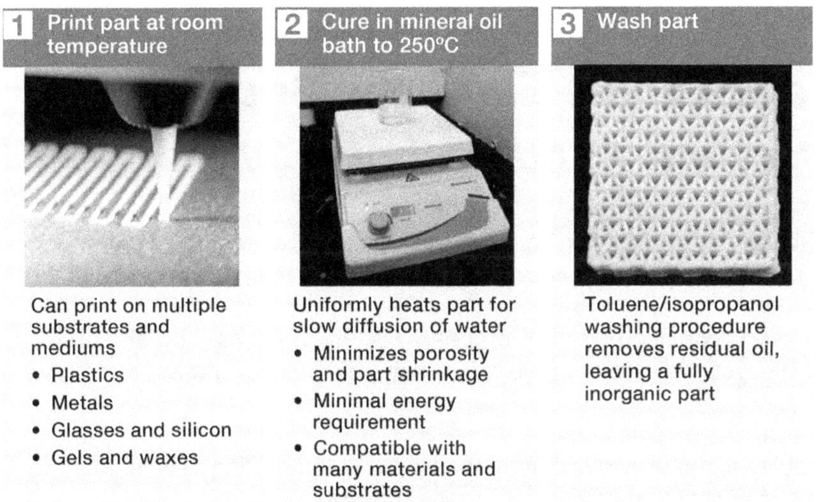

Workflow for printing glass parts

Lincoln Laboratory has demonstrated viable low-temperature 3D printing of glass structures. The process uses the direct ink writing technique to extrude, layer by layer, the Laboratory-developed multimaterial glass ink (composed of a silicate solution and nanoparticles of other inorganics) into the desired form. The process is done at room temperature. To ensure structural stability, the item is cured in a mineral oil bath heated to just 250°C. Following heat treatment, the structure is rinsed in an organic solvent to remove residual mineral, leaving a fully inorganic silica item.

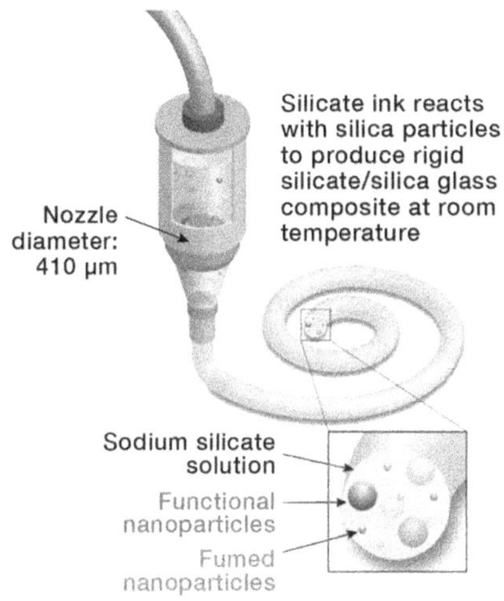

Graphic overview of custom ink engineering. Silicate ink reacts with silica particles to produce rigid silicate/silica glass composite at room temperature

Tests on items built via low-temperature additive manufacturing have shown promising results: structures with a high degree of resolution; high thermal stability, and minimal shrinkage.

Ongoing research is focused on improving the optical clarity of the glass and creating inks that enable different chemical and electrical properties.

ClearVue solar glass technology to power Australia's smartest greenhouse

ClearVue's pioneering solar greenhouse at **Murdoch University** has been operating since 2021 and has achieved remarkable results, reducing energy consumption by nearly 40%. Over four growing seasons, it has consistently produced energy and water savings, with many plant species producing higher yields in solar grow rooms.

Now the company has joined forces with **RMIT University** and 15 other organisations in an ambitious Aus$20 million research project that will position Australia as the global leader in smart, sustainable farming.

The new research programme, officially called the ARC Research Hub for Intelligent Energy Efficiency in Future Protected Cropping, represents one of the largest investments in greenhouse technology in Australian history. With Aus$5 million from the Australian Research Council (ARC) and contributions from 16 partner organisations totaling over Aus$20 million, this five-year project aims to create greenhouses that are self-sufficient and improve crop yields.

"We are excited to support this important work being led by RMIT through the ARC Hub," says Martin Deil, Global CEO of ClearVue. "Our unique technology will contribute meaningfully by generating electricity and using advanced spectral control to enhance crop yields, improve quality, and promote sustainability by reducing energy and water consumption."

Professor Gary Rothengarten, ClearVue CEO Martin Deil and :ead Scientist Dr. Mikhail Vasiliev at the Murdoch Solar Greenhouse

A booming market

The global agrivoltaics market, which combines agriculture with solar energy, was valued at $4.44 billion in 2024 and is expected to grow at a rate of 10·2% annually, reaching nearly $9.66 billion by 2032. Some projections are even more optimistic, with estimates suggesting the market could reach $7.82 billion by 2029.

This explosive growth is driven by the urgent need to feed a growing global population while reducing agriculture's environmental impact. Traditional farming consumes enormous amounts of energy and water, smart greenhouses powered by solar glass could flip that equation entirely.

What makes this solar greenhouse different?

Unlike traditional solar panels that block light, ClearVue's transparent solar glass maintains the required light transparency for optimal plants' growth while

ClearVue Technologies has joined a groundbreaking Aus$20 million research project aimed at transforming how food is grown while generating clean energy

delivering three key benefits:

- Energy Independence: Greenhouses typically consume massive amounts of electricity for lighting, heating, cooling, and automation systems. Solar glass can offset much of this demand, with some installations achieving 40% energy reductions.
- Better Growing Conditions: The spectral control technology filters light to optimize growing conditions, potentially increasing crop yields and quality. ClearVue already achieved some spectacular results at Murdoch University Solar Greenhouse, and we are looking to expand our knowledge in spectrally controlled glass for crops growth.
- Water Efficiency: The integrated systems can reduce water consumption, crucial as Australia faces increasing climate challenges.

The research consortium reads like a who's who of Australian innovation: RMIT University leads the charge with their expertise in sustainable technologies; Australian National University, University of New South Wales, and Western Sydney University bring world-class research capabilities; industry partners including South East Water Corporation, Advanced Carbon Engineering and Protected Cropping Australia provide real-world expertise; and ClearVue Technologies contributes their proven solar glass technology and hard-won experience from the Murdoch solar greenhouse trials.

Professor Gary Rosengarten from RMIT captures the vision perfectly: "Greenhouses are vital infrastructure for food security and climate resilience. By harnessing the combined expertise of our research and industry partners, RMIT is excited to be working on world leading technology like ClearVue's solar glazing to develop cutting-edge greenhouse solutions."

Australia's food security future

The five-year research programme aligns perfectly with Australia's national goals for renewable energy and food security. In a world where climate change threatens traditional farming and energy costs continue to rise, smart greenhouses represent a practical solution to multiple challenges.

"We believe Australia has an opportunity to position itself as a global leader in food security innovation, and that ClearVue should be at the forefront of this global discussion," says CEO Martin Deil.

The implications extend far beyond Australia's borders. If successful, this technology could be deployed globally, helping countries achieve food security while reducing their carbon footprint. For ClearVue, it represents a pathway to significant market expansion in the rapidly growing agrivoltaics sector.

Non-contact infrared thermometer for accurate temperature measurements in optical fibre manufacture released by LAND

LAND has developed a new non-contact infrared thermometer for precise measurement and control during deposition processes in optical fibre manufacturing.

Updating the reputable LAND Vapour Deposition Thermometer, the VDT+ is a stand-alone thermometer for use in the burner zone where the deposition takes place. Precise control of the measurement here is critical to determine the ultimate fibre strength and quality. It uses a specific wavelength in the 5 μm region where glass is opaque and interferences from atmospheric gases and products of combustion are minimal.

A wide range of highly effective mounting accessories allows ease of installation and maintenance, enabling the operator to position the thermometer in the ideal location.

The VDT+ is fully focusable with through-the-lens sighting and a narrow 100:1 field of view for precise optical alignment. A rugged and modular design ensures the pyrometer withstands harsh environments, particularly important in this application.

Temperature measurements range from 1000–2500°C and accuracy and repeatability of <1°C. The updated electronic design allows for fast, accurate, drift-free measurements to ensure a high yield. The flexible design has built-in signal processing, including signal averaging and a peak picker algorithm.

The VDT+ is a stand-alone thermometer for use in the burner zone where the deposition takes place

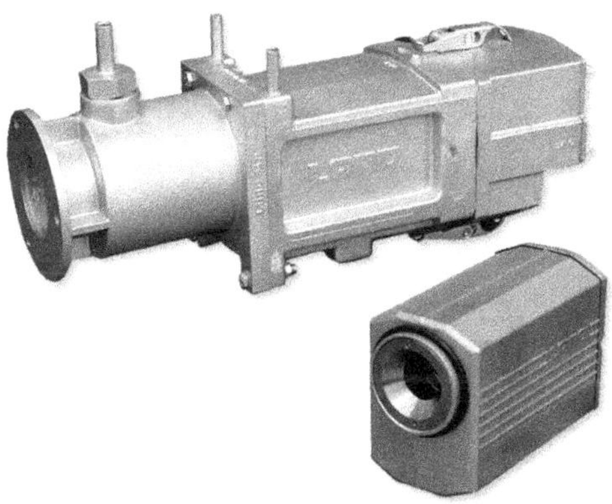

The LAND VDT+ is a stand-alone thermometer for use in the burner zone where vapour deposition takes place

Derek Stuart, Global Product Manager, from LAND said: "The VDT+ builds on our long-established VDT thermometer to deliver a powerful tool for accurate and reliable temperature measurements in optical fibre preforms and other hot glass applications. It does not require an external processor and can easily be configured using the built-in web server."

The VDT+ from LAND is the industry standard. A 4–20 mA output suits all process monitoring, recording and control systems. Individually calibrated thermometers are available with traceability to National Standards, and UKAS or NIST Calibration certificates are available.

Further details: https://www.ametek-land.com/products/non-contact-infrared-thermometers-pyrometers/vdt-vapour-deposition-thermometer

The development of MOF glasses recognised

The 2025 Royal Society of Chemistry Dalton Horizon Prize has been awarded for the discovery and development of hybrid glasses, a new family of glasses separate to known inorganic, organic and metallic families.

Hybrid glasses contain both inorganic and organic parts. They are a new category of glass, exterior to known inorganic, organic and metallic categories, and the first found since the 1970s. Applications in photonics, energy generation, gas separations, displays and memory storage may be possible.

"It is the first new chemical family of glasses formed since metallic glasses in the 1970s. The combination of organic and inorganic species within the same glass material will hopefully not only enable applications of metal-organic frameworks, but also open up new applications for glass science," said Professor Thomas Bennett.

Thomas. D. Bennett receiving the 2019 International Commission on Glass Gottardi Award from Professor John Mauro FSGT at the International Congress on Glass 2019, Boston, USA

The pioneers in this field of glasses are:

Prof. Thomas Douglas Bennett, Professor/ahorangi, School of Physical and Chemical Sciences, Te Kura Matū, University of Canterbury, Te Whare Wānanga o Waitaha, New Zealand/Aotearoa

Prof. Frédéric Blanc, Professor, University of Liverpool, UK

Prof. François-Xavier Coudert, Directeur de recherche, Chimie ParisTech, PSL University, CNRS, Institut de Recherche de Chimie Paris, France

Dr. Mathew Cowan, Senior Lecturer, MacDiarmid Institute for Advanced Materials and Nanotechnology & University of Canterbury, New Zealand

Prof. Sian Dutton, Professor, University of Cambridge, United Kingdom

Prof. Pierre Florian, Professor, CNRS, CEMHTI UPR3079, Université d'Orléans, France

Prof. Sebastian Henke, Professor, Technische Universität Dortmund, Germany

Prof. Satoshi Horike, Professor, Kyoto University, Japan & Vidyasirimedhi Institute of Science and Technology, Thailand

Assoc. Prof. Jingwei Hou, Associate Professor, The University of Queensland, Australia

Prof. Kim Jelfs, Professor, Imperial College London

Prof. David A. Keen, Professor, ISIS Facility, STFC Rutherford Appleton Laboratory, UK

Prof. Shichun Li, Professor, Institute of Chemical Materials, China Academy of Engineering Physics, Mianyang, China

Prof. John Mauro FSGT, Professor, The Pennsylvania State University United States

Prof. Sanjog Nagarkar, Professor, Indian Institute of Technology (IIT) Bombay

Prof. Morten Mattrup Smedskjær (above, left), Professor, Aalborg University, Denmark (winner of the 2018 SGT–Sir Alastair Pilkington Award.

Prof. Shane Telfer, Professor, MacDiarmid Institute of Advanced Materials and Nanotechnology & Massey University, New Zealand

Prof. Lothar Wondraczek, Professor, Otto Schott Institute of Materials Research & Center for Energy and Environmental Chemistry (CEEC Jena), University of Jena, Germany

Prof. Yuanzheng Yue FSGT (above, right), Professor, Aalborg University, Denmark

FOURTH SÃO CARLOS SCHOOL ON GLASSES AND GLASS-CERAMICS

April 27 – May 1, 2026 | São Paulo and São Carlos, Brazil

The CeRTEV team (www.certev.ufscar.br), in collaboration with ICTP-SAIFR (IFT-UNESP), is pleased to announce the *Fourth São Carlos School on Glasses and Glass-ceramics*, to be held from April 27 to May 1, 2026, following the great success of the previous editions held in 2015, 2023 and 2024. This School is part of a series organized yearly by the International Commission on Glass (ICG) in Montpellier, Wuhan and Kolkata.

SCHOOL OBJECTIVES

- Provide state-of-the-art knowledge on the structure and dynamic processes of glasses and glass-ceramics (diffusion, viscous flow, relaxation, liquid-liquid phase separation, and crystallization), as well as their optical, electrical, mechanical, and biochemical properties.
- Strengthen the international network of CeRTEV collaborators.
- Attract future students, postdoctoral researchers, and visiting scientists, while fostering collaborative research.
- Disseminate CeRTEV's faculty, infrastructure, and research facilities to Brazilian and international students and their supervisors.
- Emphasize *networking* as a key component of school.

REGISTRATION

A small registration fee of *USD 290* is required to attend the school. However, we are pleased to offer *50 grants* that will waive the registration fee, cover hotel accommodation for up to seven nights, and provide lunch expenses for selected students and postdoctoral researchers. The number of grants may increase depending on the level of industry sponsorship.

Please note that transportation, health insurance, visa fees, and any other travel-related expenses must be covered by the students themselves or by their academic supervisors or home institutions.

Young researchers from academia and industry, as well as professors, are welcome to attend the school. However, financial support is not available for these participants.

To apply for a grant, interested and qualified students and postdoctoral researchers should contact us as soon as possible on a first-come, first-served basis. Applications must be submitted by e-mail before *October 30, 2025*, and include:

- An abstract of the applicant's thesis project;
- A letter of recommendation from the thesis supervisor, including a statement confirming coverage of the applicant's transportation, health insurance, visa, and other travel-related expenses.

E-mails for applications: dedz@ufscar.br / certevlamav@gmail.com / lancelotti.r@dema.ufscar.br

Curtin's water-repellent glass breaks new ground

Curtin University researchers have developed a new technique to make glass water-repellent, a feature that could improve safety in vehicles, reduce cleaning costs for buildings and enhance filtration systems. The research, published in Advanced Functional Materials, shows how an innovative and non-toxic process using ultrasound can alter the surface of glass, making it either hydrophobic (water resistant) or electrically charged.

Lead researcher Associate Professor Nadim Darwish, an ARC Future Fellow at Curtin's School of Molecular and Life Sciences (MLS), explained that the process uses ultrasound to trigger a chemical reaction that permanently alters the surface of glass.

"The sound waves create microscopic bubbles in a diazonium salt solution, which then collapse rapidly creating tiny bursts of heat and pressure," Associate Professor Darwish said. "This triggers a reaction that forms a stable, organic layer to the glass, making it either permanently water-repellent or positively charged, depending on the type of diazonium salt used. Unlike conventional coatings, that wear off over time, our method creates a chemical bond at the molecular level, making it far more durable and environmentally friendly."

Study co-author Dr Tiexin Li, a Research Associate at Curtin's School of MLS, said the ability to modify glass surfaces in a simple and sustainable way has far-reaching implications across multiple industries.

"Glass is used everywhere—from cars and buildings to industrial filters—but its natural tendency to attract water limits its performance," Dr Li said. "Unlike traditional coatings this film won't peel off, dissolve in water or deteriorate so it's ideal for real-world applications where reliability and durability are key. This could mean clearer windshields in heavy rain, self-cleaning skyscraper windows and solar panels that stay dust-free."

Co-author Zane Datson, also from Curtin's School of MLS, highlighted another unexpected benefit, the ability of the modified glass to attract bacteria, fungi and algae.

"This is very exciting as we can tailor glass properties for specific uses including in advanced filtration systems and biofuel production," Mr Datson said. "For example, the coated glass can help bind yeast in brewing, capture bacteria in wastewater filtration systems or act as a chemical barrier to micro-organisms in air filters."

The research team is now seeking industry partners to test and scale up the technology, particularly in the automotive, construction and environmental sectors.

The full paper titled, 'Sonochemical Functionalization of Glass,' can be found online

Unlocking new optical power in glass nanostructures using precision printing

Researchers created nanoscale glass structures with near-perfect reflectance, overturning long-held assumptions about what low-index materials can do in photonics. Glass has been a reliable component of optical systems, valued for its transparency and stability, however at the nanoscale, especially for high-performance optical devices, glass has traditionally taken a backseat to higher refractive index materials. Now, a research group led by Professor Joel Yang from the Singapore University of Technology and Design (SUTD) has addressed this issue.

With findings published in *Science Advances*, the group has developed a new method to 3D-print glass structures with nanoscale precision and achieve close to 100% reflectance in the visible spectrum, opening up a broader role for glass in nanophotonics, including in wearable optics, integrated displays, and sensors.

The researchers' breakthrough is enabled by a new material called Glass-Nano: a photocurable resin made by blending silicon-containing molecules with other light-sensitive organic compounds. Unlike conventional approaches that rely on silica nanoparticles, Glass-Nano cures smoothly and contracts uniformly during heating, transforming into clear, robust glass. When printed using two-photon lithography, these polymer structures shrink during sintering at 650°C, preserving their form while achieving nanoscale features as small as 260 nm.

"Instead of starting with silica particles, we worked with silicon-bearing molecules in the resin formulation," explained Prof Yang. "This resin enables us to build up nanostructures with much finer detail and smoother surfaces than was previously possible. We then convert them into glass using our 'print-and-shrink' process without sacrificing fidelity."

The team focussed their fabrication on photonic crystals (PhCs), artificially structured materials featuring repeating patterns that interact with specific wavelengths of light. These structures can reflect light very efficiently, but only if built with extreme regularity and precision. Previous efforts to realise low-index 3D PhCs have consistently fallen short, exhibiting only poor reflectance due to structural irregularities and distortions.

"This resin enables us to build up nanostructures with much finer detail and smoother surfaces than was previously possible. We then convert them into glass using our 'print-and-shrink' process without sacrificing fidelity," says Professor Joel Yang, Singapore University of Technology and Design.

With their new method, the researchers overcame these limitations. By printing more than 20 tightly stacked layers and fine-tuning the design geometry, they achieved a structurally highly uniform, diamond-like photonic crystal that reflects nearly 100% of incident light within a broad range of viewing angles.

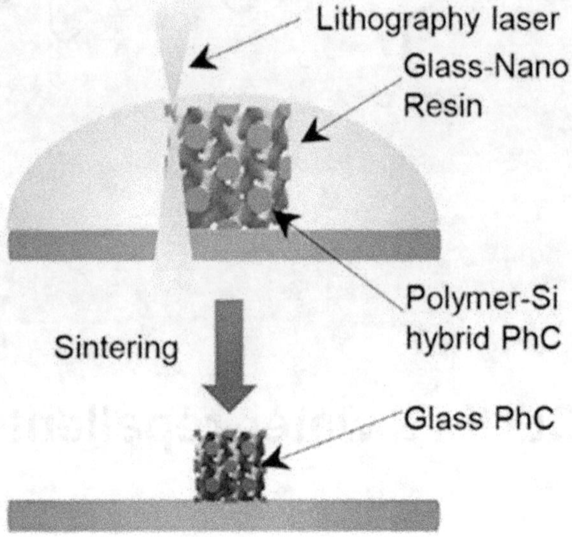

Schematic of the printing and sintering process. TPL was used to polymerise the resin to 3D microstructure. After printing, the structure was heated in air to remove the organic composition, and oxygen reacts with silicon within the structure to form silica

"Historically, low-index materials like silica were seen as optically weak for this purpose. But our findings show that with enough uniformity and structural control, they can outperform expectations—and even rival high-index materials in reflectance," says Dr Wang Zhang, SUTD Research Fellow and first author of the paper. "At the macroscale, shrinkage like this would collapse the structure. But at the nanoscale, the high surface-to-volume ratio actually helps preserve stability. Our resin formulation, engineered with multiple cross-linkers and a silicon-rich precursor, ensures both the printability and the mechanical robustness needed to survive the heat treatment."

"Even tiny spectral reflectance features—so small that we originally suspected they might be measurement artifacts—line up well with calculated predictions of standing-wave oscillations," said Associate Professor Thomas Christensen, a co-author of the paper from the Department of Electrical and Photonics Engineering at the Technical University of Denmark.

"With the ability to print high-resolution nanostructures in both low- and high-index dielectrics, we're now turning to applications where 3D optical components could reduce transmission losses and enable more efficient photonic systems," said Prof Yang.

Alfred University Professor Emeritus L. David Pye Hon FSGT has been awarded the honorary degree of Doctor of Humane Letters. Pye received his bachelor's and doctoral degrees in ceramic science from Alfred University in 1959 and 1968. He joined the faculty of the New York State College of Ceramics (NYSCC) at Alfred University in 1968, beginning an academic career that lasted until 2002 and included serving as dean of NYSCC.

Kathleen Richardson FSGT, the first woman to receive a PhD in glass science from Alfred University, in 1992, read the citation praising Pye's work as a world leader in ceramic technology and research, citing "his contributions to glass and ceramic science at Alfred University and the world."

Professor Richardson was joined by Alfred University President Mark Zupan, Gabrielle Gaustad, dean of the Inamori School of Engineering and vice president for Statutory Affairs, Lauren Lake, dean of the School of Art & Design, and Brian Sullivan, dean of University Libraries. Gaustad recalled Pye supporting her husband, and countless other NYSCC students, observing he "always had time to meet with students." Zupan cited Pye's generous work for the AU community, comparing him with the Alfred jewel in the University mace. "Our true jewels are our graduates such as David," Zupan said.

Pye's accomplishments in the ceramic science community are legendary. They include his election as the 46th Honorary Fellow of the Society of Glass Technology and his award of a Distingished Life Membership in the American Ceramic Society (ACerS) for lifetime achievements in the field of ceramics. In 2019, the Glass and Optical Materials Division of ACerS established the L. David Pye Lifetime Achievement Award, and he served as Founding Editor of ACerS' International Journal of Applied Glass Science. David Pye also served as President of the International Commission on Glass.

In his own remarks after receiving the honorary degree, Pye thanked his family, whose many members were in attendance, and quoted from Lou Gehrig's 1939 farewell speech at Yankee Stadium: "Today, I consider myself the luckiest man on the face of the earth."

Society of Glass Technology Meeting on History and Heritage

The Ebb and Flow of Glass Compositions

3rd September 2025

At Murray Edwards College, Cambridge

The Society of Glass Technology is planning a one day event during its Annual Meeting at Murray Edwards College on 3rd September 2025. The £65 fee includes lunch, coffee and tea breaks. Limited places are also available at the conference banquet on 2nd September. Booking is via the SGT conference **web site at this link: https://sgt.org/**

We are seeking presentations on Glass History and Heritage, particularly on glass composition. Abstracts of up to 200 words should be submitted at: www.sgt.org. While 20 minute slots will be the norm please indicate if you are willing to speak for 40 minutes.

A team of judges will consider all presentations for the David Martlew Award unless authors request otherwise. The winner will be invited to attend the H&H session of the 2026 Annual Conference along with the 2026 conference banquet free of charge, to receive formal recognition for their award.

Glass Artefact compositions have changed over time and location according to raw material availability, melting characteristics and the required product properties. Over recent decades chemical analysis techniques have improved dramatically so Archaeologists and Historians can now track these changes with remarkable precision, leading to deeper understanding of glass production and the factors influencing it.

We hope our event will encourage the glass community. including glass technologists, document researchers, experimental-glass makers, analysts of atomic spectra, specialist glass analysts, and customers for the information generated such as glass collectors etc., to engage in conversations on glass compositions and what they tell us of the recent history of glass artefacts and their manufacture.

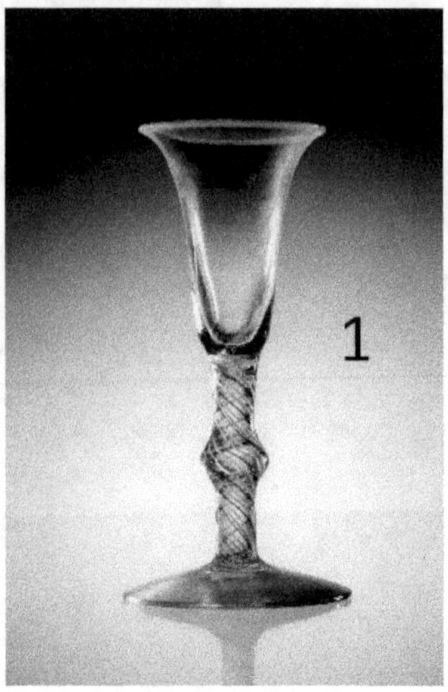

As a second related thread, we invite presentations that examine the consequences for contemporary glass artists around questions of sustainability, and the responsible production and consumption of glass. For example, we invite talks that detail the use of different colorants, different raw materials and cullet that lower temperatures for glass making, the use of recycled glasses, the use of low impact alternatives for processing glass and the availability of different heat sources such as hydrogen, biofuels and electricity to fuel glass making.

Society of Glass Technology
Annual Meeting 2025
Murray Edwards College, Cambridge

1–3 September 2025

Programme
and
Abstracts

Sponsored by

AGC

and

ŞİŞECAM

YEARS

Society of Glass Technology

9 Churchill Way
Chapeltown
Sheffield S35 2PY, UK
Tel +44(0)114 263 4455
Email info@sgt.org
Web www.sgt.org

The Long Room

09.00	**Welcome** Angela Seddon FSGT, SGT President
09.10	**N. L. Varshneya Memorial Lecture:** Glass sustainability: challenges and opportunities at end of life Gabrielle Gaustad, Alfred University
09.50	Boron coordination changes with temperature in a barium borosilicate glass and melt using neutron diffraction Oliver L. G. Alderman, ISIS Facility Rutherford Appleton Laboratory
10.10	The principle of maximal connectivity applied to borosilicate glasses Richard Kerner, Sorbonne-University
10.30	Thermodynamics and kinetics of ion exchange in silicate glasses Guglielmo Macrelli, Isoclima
10.50	Effect of the local structure around transition elements on their properties in oxide glasses L. Galoisy, Sorbonne University
11.10	**Coffee**
11.40	**2025 Darshana and Arun Varshneya Frontiers of Glass Lecture:** What use is a really bad glass? S. R. Elliott, University of Oxford
12.20	Structure of lithium alumino-titano-phosphate $Li_2O–Al_2O_3–TiO_2–P_2O_5$ (LATP) glasses as possible electrolytes for lithium batteries Gavin Mountjoy, University of Kent
12.40	Dissolution of phosphate glass examined through computer simulation using reactive force fields Jamieson K. Christie, Loughborough University
13.00	**Lunch**

CAMBRIDGE 2025 : MONDAY 1 SEPTEMBER : SCIENCE

14.00	**SGT–Sir Alastair Pilkington Award:** Combining machine learning, simulations and experiments for improved glass design N. M. Anoop Krishnan, Indian Institute of Technology Delhi
14.40	Clustering of iron in glasses: composition–structure–property relations Paul A. Bingham, Sheffield Hallam University
15.00	Molecular dynamics simulations of the environment of alkali ions in mixed lithium-sodium silicates Alastair N. Cormack, Alfred University
15.20	Johari–Goldstein relaxation in rejuvenated chalcogenide glasses Jacopo Baglioni, University of Padova
15.40	**Coffee**
16.00	Bond angles and the structure of vitreous BeF_2 Alex C. Hannon, ISIS Facility Rutherford Appleton Laboratory
16.40	Where does sodium (Na) go in sodium silicate glasses? Implications for physical properties and behaviour Grant S. Henderson, University of Toronto
17.00	Physical properties of sodium germanate glasses over an extended range of compositions Zachary Pearlman, Coe College
17.20	Structure and properties of high-density lead bismuth borosilicate glasses Brandon Herrejon-Zavalza, Coe College
17.40	Characterisation by ^{31}P NMR spectroscopy of acid glass of potassium phosphate Fatma Zahra Souissi, University of Sfax and Gabes University
18.00	From melt to glass stabilised below the glass transition temperature: structural changes Natalia M. Vedishcheva Institute of Silicate Chemistry, St. Petersburg
18.20	**Close**

CAMBRIDGE 2025 : MONDAY 1 SEPTEMBER : SUSTAINABILITY

	The Long Room then Vivien Stewart Room from 09.50
09.00	**Welcome** Angela Seddon FSGT, SGT President
09.10	**N. L. Varshneya Memorial Lecture:** Glass sustainability: challenges and opportunities at end of life Gabrielle Gaustad,Alfred University
09.50	Assessment of irreversibility in the performance of glass furnaces R. Conradt, uniglassAC
10.30	How has our industry evolved over the last year Steve Whettingsteel, Krysteline Technologies
11.10	**Coffee**
	Vivien Stewart Room
11.40	
12.10	The practicalities of implementing a carbon strategy Franck Crosnier, Edenseven
12.40	Powered by people: the path to net zero for glass is not just a technical challenge it is a human one! Linda Newnes, University of Bath
13.00	**Lunch**

CAMBRIDGE 2025 : MONDAY 1 SEPTEMBER : SUSTAINABILITY

	Vivien Stewart Room
14.00	Overview of CO_2 capture systems with hot potassium carbonate (HPC) applied to glass furnaces Giorgio di Lascio, K2-CO2
14.30	Twenty years of experience in yhe valorization of glass pozzolan in concrete Arezki Tagnit-Hamou, University of Sherbrooke
15.40	**Coffee**
16.00	Cullet, glass fibres etc Gareth Morris, Saint Gobain
16.40	Challenges and progress in recycling of specialty glass and glass-ceramics Jörn Besinger, SCHOTT
17.10	Glass furnace lining recycling a challenge for sustainable refractory industry Tiphane Bazin, Valoref
17.40	How the future of glass production could look Stuart Hakes, F.I.C. (UK)
18.10	**Close**

CAMBRIDGE 2025 : TUESDAY 2 SEPTEMBER : SCIENCE

The Long Room

09.00 On the role of structural changes at the glass–liquid transition
Michael Ojovan, University of Sheffield

09.40 Analysis for boron using the "OXI" XRF standard analysis method at Sheffield Hallam University
Anthony Bell, Sheffield Hallam University

10.00 Vitrification of uranium tetrafluoride
Mike T. Harrison, United Kingdom National Nuclear Laboratory

10.20 Behaviour and treatment of electrorefiner salt simulants during nuclear fuel processing
Lucas Greiner, Missouri University of Science and Technology

10.40 The vitrification of waste containing metallic uranium from the nuclear industry
Jessica L. Blenkinsop, United Kingdom National Nuclear Laboratory

11.00 Coffee

11.20 Femtosecond laser induced plasma modification of silicate glass
Gin Jose, University of Leeds

12.00 The effect of impurities and irradiation on the glass network – Cs_2O-loaded iron phosphate case study
Cillian Cockrell, Bangor University

12.20 On the thermal and kinetic methods of nepheline precipitation within Hanford high-level nuclear waste canisters
Jack D. Cornwall, Sheffield Hallam University

12.40 Thermal treatment of radiological asbestos-containing wastes (ACW): a disposability toolbox
Hannah E. Blystra, Sheffield Hallam University

13.00 Lunch

CAMBRIDGE 2025 : TUESDAY 2 SEPTEMBER : SCIENCE

The Long Room

14.00 **Plenary Lecture:** Rare earth optical materials
Setsuhisa Tanabe, Kyoto University

14.40 Integration of rare earth activated phosphor into inorganic glass: from bulk phosphor in glass to phosphor-glass thin films
Jeena Rose Jose, University of Leeds and Mahatma Gandhi University

15.00 Tailoring structural and functional properties of phosphate glasses through CaO and Al_2O_3 incorporation
Yasmina Alaoui, Mohammed V University in Rabat

15.20 Photo-induced PL colour change and anion migration of $CsPb(Br,I)_3$ perovskite nanocrystals in glass-ceramics
Kejie Huang, Kyoto University

15.40 Coffee

16.00 **Plenary Lecture:** A journey from soft glasses to photonics devices
Yasutake Ohishi, Toyota Technological Institute

16.40 Development of a structural model of strontium doped bio-phosphates for the interrogation of medium-range order
Daniel Bradley

17.00 Rhenium volatilization from molten glass: relationship between volatilization and viscosity
Chenchen Niu/Kai Xu, Wuhan University of Technology

17.20 Switching and thermal behaviour of Cu and Sn doped Si–Te glasses: a comparative analysis
Chandasree Das, BMS College of Engineering and Visvesvaraya Technological University

17.40 Structural effects of controlled iron redox in obsidian-like glass systems
Andrew Parnapy, Alfred University

17.40 End
– Followed by Reception (18.45) and Conference Banquet (19.30) at Magdalene College

CAMBRIDGE 2025 : TUESDAY 2 SEPTEMBER : SUSTAINABILITY

	Vivien Stewart Room
09.00	Foam Glass manufacturing as sustainable reuse of waste glass Arjen Steiner, Foamit
09.30	A producer's responsibility and strategy for carbon reduction Graeme DeBrincat, ARUP
10.00	AGC's Volta decarbonisation of flat glass melting Project Niels Schreuder, AGC
10.30	How technology is shaping the recovery of cullet Vasco Sousa, PICVISA
11.00	**Coffee**
11.20	Challenges and early successes of running a 30 tonnes per day pilot line Bridget Stewart, Glass Futures
11.40	The next generation of industrial glass furnaces Owen McGann, Glass Technology Services (GTS)
12.00	Beneficiation of ash and slag for sustainable glass production Sabrin A. Samad, Sheffield Hallam University
12.20	Reducing energy consumption and carbon dioxide emissions from glass manufacture Ebele Ahizi, Sheffield Hallam University
12.40	Vitreous radioactive wasteform development: an overview Rachel Crawford, Sheffield Hallam University
13.00	**Lunch**

CAMBRIDGE 2025 : TUESDAY 2 SEPTEMBER : SUSTAINABILITY

	Plenary Lectures in the Long Room then Sustainabilty in the Vivien Stewart Room
14.00	**Plenary Lecture:** Rare earth optical materials Setsuhisa Tanabe, Kyoto University
14.40	Sodalime and volcanic ash alkali activated foams for organic dye sorption G. Tameni, University of Padova
15.00	A bridge between industry and academia Rob Werner, Glass Technology Services (GTS)
15.40	**Coffee**
16.00	**Plenary Lecture:** A journey from soft glasses to photonics devices Yasutake Ohishi, Toyota Technological Institute
16.40	Summarising the SGT Sustainability Sessions and formulating a plan for a low carbon future Katrina Skerratt-Love & Collin Wilkinson, University of Liverpool & Alfred University
17.20	Murano glass awards for the contributors who excelled during the event Krysteline, Krysteline Technologies
17.40	**End** **– Followed by Reception (18.45) and Conference Banquet (19.30) at Magdalene College**

CAMBRIDGE 2025 : WEDNESDAY 3 SEPTEMBER : SCIENCE

The Long Room

09.00	**Plenary Lecture:** Mid-infrared photonics: a new frontier in disease screening and diagnosis using chalcogenide glass fibreoptics Angela B. Seddon, University of Nottingham
09.40	Optimisation of Er^{3+}/Yb^{3+} co-doped phosphate-based laser glasses for enhanced laser performance Ahmet Caner Kayaalp, Şişecam Science, Technology and Design Center
10.00	"The Salts of the Rare Metals": a physical-chemical description of Frederick Carder's aurene glass Annika Blake-Howland, Alfred University
10.20	Structure and redox state of the sulphur-containing glass melts Justyna Sułowska, AGH University of Krakow
10.40	**Coffee**
11.00	Hollow core fibres: when nothing is better than glass! Francesco Poletti, University of Southampton & Microsoft Azure Fiber
11.40	Tailoring rapid dissolution of metaphosphate glasses for biocompatible applications Lucas Greiner, Alfred University and Missouri University of Science and Technology
12.00	Advances in glass-based space radiation shielding Katrina Skerratt-Love, University of Liverpool
12.20	pH switchable bacteriocidal glasses Robert Hill, Queen Mary University of London
12.40	At the intersection of art and science: exploring dichroic glass through reconstructions to use as an artistic medium Zoe Bedford, Independent researcher, glass artist
13.00	**Lunch**

CAMBRIDGE 2025 : WEDNESDAY 3 SEPTEMBER : SCIENCE

The Long Room

14.00	High aspect ratio flat optical fibre for application in process control evaluation and health monitoring of high value composite assets Christopher Holmes, University of Southampton
14.40	Quantification of chemical replacement zones in glasses by means of ToF-SIMS, NRA and nanoindentation Karl-Michael Weitzel, Philipps-Universität Marburg
15.00	New MWIR (mid-wave infrared) fibre lasers for sensing Thomas Readyhoof, University of Nottingham
15.20	**Coffee**
15.40	Sculpting glass with light for future fibre optics Calum A. Ross, Heriot-Watt University
16.20	Manifestation of the mixed alkali effect on viscosity in the $K_2O/Na_2O–CaO–SiO_2$ system Wei Deng, Sheffield Hallam University
16.40	Tunable luminescence and realisation of warm white light via energy transfer from $Dy^{3+}/Er^{3+}/Sm^{3+}$ triply doped glasses P. R. Biju, Mahatma Gandhi University
17.00	Eu^{3+}/Dy^{3+} co-activated multicomponent borosilicate matrix: multifunctional glass for lighting and solar energy harvesting Adon Jose, St. Stephen's College Uzhavoor
17.20	Multicomponent glasses for hollow core antiresonant fibres Mahmadur Rahman, University of Southampton
17.40	**Close**

CAMBRIDGE 2025 : WEDNESDAY 3 SEPTEMBER : HERITAGE & HISTORY

The Long Room then from 09.40 Vivien Stewart Room 1

09.00	**Plenary Lecture:** Mid-infrared photonics: a new frontier in disease screening and diagnosis using chalcogenide glass fibreoptics Angela B. Seddon, University of Nottingham
09.40	The evolving future of sustainable glass art practice Jessamy Kelly, Edinburgh College of Art
10.00	Re-contextualising scientific glass: unlocking alternative perceptions of value in high jewellery Estelle Burton
10.20	Yellow colour of stained-glass windows: a French historical perspective Georges Calas
10.40	**Coffee**
11.00	"The Salts of the Rare Metals": a physical-chemical description of Frederick Carder's aurene glass Annika Blake-Howland, Alfred University
11.20	The charm of combining glass with other materials in artistic creation Jianyong Guo
11.40	Glass & Light: materialities & metaphors Charlie Murphy
12.00	Investigation on the contemporary development status of traditional glass crafts in Boshan of China Dian Shi
12.20	Prototyping Creative Circularity in Luxury Tableware Lulu Harrison, Lulu Harrison Studio, Oxfordshire, UK
12.40	Cycles of transformation: exploring lifecycles in glass art making and material reuse Gregory Alliss, Edinburgh College of Art
13.00	**Lunch**

CAMBRIDGE 2025 : WEDNESDAY 3 SEPTEMBER : HERITAGE & HISTORY

Vivien Stewart Room 1

14.00	Recovering Secrets from Historic Glass-waste Colin Brain
14.40	An incomplete history of didymium for glassblowing Andrea Sella, University College London
15.00	Imitating the Lycurgus Cup Lisa Hollands, University of Sheffield
15.20	**Coffee**
15.40	Six of One, and Half-a-Dozen of the Other Clive Manison
16.00	Is glass composition a unique fingerprint for a 19th century glass maker? John Parker, University of Sheffield
16.20	Bring and Show Colin Brain
17.20	**Close**

	Vivien Stewart Room 2
	UNLOCKING THE MYSTERIES OF GLASS
09.00	Part I Fundamentals of Glass Science & Technology Topics covered: Families of commercial glasses; What is glass? Glass structures; Metastable phase separation; Viscosity; Glass transition range behaviour; Chemical durability; Hardness, Strength and Strengthening; Q&A Instructor: Professor Arun Varshneya, Alfred University, USA
13.00	**Lunch**
	Vivien Stewart Room 2
	UNLOCKING THE MYSTERIES OF GLASS
14.00	Part II: Fractographic analysis of glass products Topics covered: Basics of glass fractographic analysis. Hands on examination of glass fractures. Instructor: Daniel Capon, Technical Manager, Glass Technology Services, Sheffield.
17.00	**Close**

Phys. Chem. Glasses: Eur. J. Glass Sci. Technol. B, 2025, **66**(4), 113 DOI: 10.13036/17533562.66.4.AM01

Glass sustainability: challenges and opportunities at end of life

Gabrielle Gaustad, Collin Wilkinson, Deniz Saadatpour, Arron Potter, River Pao, Scott Misture, Rebecca Welch, Greg Palmer, William LaCourse, Alexis Clare, Elizabeth Tsekrekas, Charles Bellows, Stuart Yaniger & Lucas Greiner

Alfred University, Kazuo Inamori School of Engineering, New York State College of Ceramics, 1 Saxon Drive, Alfred, NY 14802, USA

Corresponding author email: gaustad@alfred.edu

Glass recycling economics are a growing challenge for US Material Recovery Facilities (MRFs). Most collection has turned to single stream to reduce costs and transportation distances; this creates a glass product from municipal solid waste that is highly contaminated, or has significant costs to separate and purify. Glass, theoretically, is completely recyclable but while 12 million tons of glass are produced each year in the United States, experts estimate that less than 25% is recycled (Jacoby 2019). The low recycling rate in the US may surprise some, as the European Union has container glass that consists of 80-90% to recycled cullet compared to only 30-40% in the US (Cooper and Swiler 2019). Use of recycled cullet reduces the amount of post-consumer glass going into landfills and also can reduce energy use of producing glass by up to 20% (Dyer 2014) also saving 5% of greenhouse gas emissions for every 10% of cullet substituted for primary glass (Kovacec, Pilipovic et al. 2011).

This presentation highlights work from an Alfred University and New York State Department of Environmental Conservation partnership exploring the potential for improving reuse and recycling rates of glass in the US. Projects explore both technologies for improving quality and identifying sinks for lower quality recycled glasses including use in a wide range of sinks including agricultural applications, healthcare, cement and concrete production including novel 3D printing approaches, and asphalt and road base. The potential for glass bottle reuse in specific regions as well as sorting and separation technologies will also be detailed. Broader modelling efforts to understand the impacts of policy decision in New York State ...

Phys. Chem. Glasses: Eur. J. Glass Sci. Technol. B, 2025, 66(4), 114 DOI: 10.13036/17533562.66.4.AM02

Boron coordination changes with temperature in a barium borosilicate glass and melt using neutron diffraction

Oliver L. G. Alderman

ISIS Facility Rutherford Appleton Laboratory, Chilton, Didcot, Oxon OX11 0QX, UK

Corresponding author email: oliver.alderman@stfc.ac.uk

Barium borosilicate glasses are of applied interest for nuclear waste immobilization, waste water treatment and as optical materials. They are also well suited to high-temperature hermetic sealing applications, including for solid oxide fuel cells, where both the properties of the glasses and the supercooled liquids are of importance. Borate bearing liquids have been shown to have anomalously large thermal expansion coefficients compared to their glasses,[1] *making them of renewed fundamental interest, and this is presumably related to the fact that the tetrahedral boron fraction is known to vary strongly with temperature for many (but not all) compositions.*[2–4] *Herein we present a neutron diffraction study using the SANDALS instrument at the ISIS Neutron and Muon Source, UK, on a barium borosilicate glass and its supercooled liquid up to T=1200 K, showing clear evidence for a tetrahedral to trigonal boron transition above the glass transition temperature (c. 900 K). We discuss the interpretation of the data in terms of simplified chemical equilibria, compare to similar measurements on pure* $^{11}B_2O_3$,[5] *and discuss the challenges and opportunities presented by these types of experiments.*

References

1. P. Lunkenheimer *et al.*, *Nat. Phys.* **19**, 694 (2023).
2. O. L. G. Alderman *et al.*, *Rep. Prog. Phys.* (2025) **DOI** 10.1088/1361-6633/adc69c.
3. O. L. G. Alderman *et al.*, *Int. J. Appl. Glass Sci.* **14** (1), 52 (2023).
4. O. L. G. Alderman *et al.*, *J. Chem. Phys.* **158** (22), 224501 (2023).
5. O. L. G. Alderman, *J. Chem. Phys.* **162** (5), 054502 (2025).

Phys. Chem. Glasses: Eur. J. Glass Sci. Technol. B, 2025, **66**(4), 115 DOI: 10.13036/17533562.66.4.AM03

The principle of maximal connectivity applied to borosilicate glasses

Richard Kerner

LPTMC, Sorbonne-University, 4 Place Jussieu, 75005 Paris, France

Corresponding author email: richard.kerner@sorbonne-universite.fr

The influence of alkali modifiers (Na_2O, K_2O) on the glass transition temperature is well known. Their presence leads to an opposite effect in alkali borates $(M_2O)_x(B_2O_3)_{(1-x)}$ where $T_g(x)$ grows with x, and in alkali silicates $(M_2O)_x(SiO_2)_{(1-x)}$ for which $T_g(x)$ is rapidly decreasing when x is growing, the principal reason being the change of the average coordination number due to the creation of four-coordinate boron atoms in the first case, and the three- or even two-coordinate silicon atoms in the second. Theoretical models based on Maxwell's rigidity count and the so-called "floppy modes" were developed by J.C. Phillips & M. Thorpe in the 1980s,(1) followed by more elaborated ones proposed by us in the 1990s (R.K.(2) & M. Micoulaut, R.A. Barrio, D.-M dos Santos, G. Naumis and others), successful in describing covalent glasses and alkali-borates and sulphites. A more sophisticated model of alkali borates was introduced recently, based on principles of maximal homogeneity and conservation of the connectivity of the network when the alkali modifier's content is increasing.

Borosilicate glasses present a challenge for a similar model building, because two tendencies, increasing and decreasing coordination number, oppose each other, thus making more difficult the derivation of a system of differential equations describing the average behavior of essential parameters besides the coordination number, characterising medium-range order with local configurations count, the statistic of rings and average rigidity. We shall present the model of borosilicate glasses with low alkali content, enabling the evaluation of glass transition temperature behavior, rigidity and elasticity coeficients.

References

1. Structure and function of window glass and Pyrex , J C Phillips, R Kerner, *J Chem Phys* 2008 May 7;128(17):174506. doi: 10.1063/1.2805043.
2. Evolution of local structures in alkali-borate glasses, Richard Kerner, Dina Maria dos Santos-Loff & Ana Cristina Rosa, *Physics and Chemistry of Glasses: European Journal of Glass Science and Technology Part B*, DOI : 10.13036/17533562.65.2.16

Phys. Chem. Glasses: Eur. J. Glass Sci. Technol. B, 2025, **66**(4), 116 DOI: 10.13036/17533562.66.4.AM04

Thermodynamics and kinetics of ion exchange in silicate glasses

Guglielmo Macrelli,[1] *Arun K. Varshneya*[2] *& John C. Mauro*[3]

[1] *Isoclima SpA, R&D Department, Italy*
[2] *Saxon Glass Technologies Inc. & Alfred University, Alfred, NY, USA*
[3] *Department of Materials Science and Engineering, The Pennsylvania State University, University Park, PA 16802, Pennsylvania, USA*

Corresponding author email: guglielmo.macrelli@isoclimagroup.com

Ion exchange is a well-established technology in silicate glass where the underpinning physics is widely known,(1) *although some particular issues are still debated in the scientific glass community.*(2) *Ion exchange is a kinetic process that occurs in non-equilibrium systems.*(3) *A thermodynamic analysis of ion exchange based on thermodynamics of open subsystems helps in identifying the equilibrium conditions and the driving forces that activate the mutual exchange of alkali ions from a silicate glass to a molten salt reservoir. The thermodynamic approach cannot detail how the incoming ions concentration evolves in the glass but can identify suitable parameters controlling equilibrium conditions at the glass/bath interface. The establishment of surface equilibrium on a time scale much shorter than the overall immersion time helps in establishing suitable boundary conditions to solve the interdiffusion kinetic problem. A simple, full kinetic model is proposed for the description of ion exchange where a key role is played by the introduced surface saturation time at glass surface and its ratio to the overall immersion time. Further discussion is focused on a thermodynamic factor that links thermodynamics and the interdiffusion coefficient and, when it is different from unity, represents the non-ideal behavior of the solution of the silicate glass with respect to the incoming ions. A new expression for the interdiffusion coefficient, derived from the thermodynamics of irreversible processes,*(4) *is proposed. Based on this new structure of the interdiffusion coefficient, a discussion is presented to explain the contribution to the interdiffusion coefficient reported in literature*(5) *coming from the interaction effects of the silica network with the exchanging ions.*

References

1. Varshneya, AK. & Mauro, JC, *Fundamentals of Inorganic Glasses 3rd Edition,*. Elsevier, Amsterdam, 2019.
2. Varshneya, AK, *Int.J.Appl.Glass Sci.*1[2], 131-142, 2010
3. Mauro, JC, *Material Kinetics, Transport and Rate Phenomena*, Elsevier, Amsterdam, 2021.
4. Macrelli G, arXiv.2401.10916v6. https://doi.org/10.48550/arXiv.2401.10916, 2024.
5. Varshneya, AK, *J.Am.Ceram.Soc.*,58(3-4);106-109, 1975

Phys. Chem. Glasses: Eur. J. Glass Sci. Technol. B, 2025, **66**(4), 117 DOI: 10.13036/17533562.66.4.AM05

Effect of the local structure around transition elements on their properties in oxide glasses

L. Galoisy, G. Calas, L. Cormier & G. Lelong

Institut de Minéralogie, de Physique des Matériaux et de Cosmochimie (IMPMC), Sorbonne Université, UMR CNRS 7590, MNHN – 4 place Jussieu, 75005 Paris, France

Corresponding author email: laurence.galoisy@sorbonne-universite.fr

The structural environment of cations in oxide glasses is of prime importance as it is linked to glasses fundamental properties such as durability, glass alteration or reactivity. The local environment of different cations has been investigated at a short and medium range scale in multicomponent silicate, aluminosilicate, borate and borosilicate glasses, using spectroscopic methods, scanning electron microscopy (SEM) and radiation diffraction. These data give either a description of the geometry and symmetry of the cationic site, including the nature of the chemical bond. Cations exhibit several original structural properties in oxide glasses. Transition elements such as Ni or Ti may occur in 5-coordinated sites which can coexist with other coordination numbers, depending on glass composition. The environment of cations such as Zn, Zr or Mo is discussed using the bond valence theory, which predicts the way to charge compensate the oxygen neighbours and which indicates the linkage of cationic sites with the silicate framework giving specific structural role to the cations. Oxide glasses can also exhibit elements with various oxidation states and coordination sites, as uranium giving for instance, different colours to the glass. The determination of the sites occupied by the elements in their different oxidation states allows also to rationalise the chemical dependence of redox equilibria, which is one way to predict Fe behaviour in magmatic silicate systems. Several experimental data lend support that cations are located in domains extending up to more than 8 Å radius, in which cationic polyhedra may be linked together either by edges or by corners. In low alkali borate and borosilicate glasses, transition elements such as Co, Ni, Zn exist in peculiar highly ordered domains corresponding to the presence of rigid borate units.

These results show that strong differences exist between modifying and charge compensating cations, either concerning site geometry or medium range organisation, in relation with their peculiar properties.

Phys. Chem. Glasses: Eur. J. Glass Sci. Technol. B, 2025, **66**(4), 118 DOI: 10.13036/17533562.66.4.AM06

What use is a really bad glass?

S. R. Elliott

Physical and Theoretical Chemistry Laboratory, University of Oxford, Oxford, UK

Corresponding author email: stephen.elliott@chem.ox.ac.uk

In conventional glass science and technology, the thermal stability of a glass is often of primary importance. What then, rather counter-intuitively, might be the technological applications, if any, of glasses exhibiting extremely bad thermal-stability characteristics, i.e. with exceedingly fast crystallization kinetics?

In this lecture, I will discuss such glassy materials, principally tellurides alloyed with metalloid elements, such as germanium or antimony, which exhibit crystallization times of a few nanoseconds. In other words, these must be among the worst possible glass formers from a thermal-stability viewpoint. However, the fact that such materials have large opto-electronic property contrasts between the metastable crystalline phase (electrically-conducting 'SET' state, {1}) and the glassy phase (electrically-resistive 'RESET' state, {0}) means that they can form the basis of new non-volatile electronic (or optical) solid-state memory devices – so-called 'phase-change memory' (PCM). Binary bits of information {0,1} are stored as a metastable structural state of the memory material, rather than as electrons trapped at the floating gate in conventional silicon CMOS flash-memory devices.

I will discuss our recent million-atom, device-scale molecular-dynamics (MD) simulations of SET and RESET processes in Ge-Sb-Te PCMs using linear-scaling, machine-learned (ML) interatomic potentials, having density-functional-theory (DFT) levels of accuracy. I will present experimental results for a new type of Te-based selector switch for use in cross-point PCM arrays, where the OFF state is the semiconducting crystalline phase and the ON state is the electrical-pulse-generated metallic liquid state. ML-MD simulations of Te reveal the origin of the decrease in threshold switching voltage to be due to preferential melting of the polycrystalline material at grain boundaries. In addition, experimental evidence reveals that, intriguingly, the $GeTe_x$ *system exhibits three distinct types of switching/memory behaviour, depending on the composition: (i) non-volatile crystal-glass-crystal PCM behaviour for x=1 (GeTe); (ii) volatile all-glass electronic-switching 'Ovshinsky-threshold-switch' (OTS) behaviour for x=2–7; and (iii) volatile crystal–liquid–crystal 'phase-change-switch' (PCS) behaviour for compositions for x=8, 16, and pure Te. This behaviour can be rationalised in terms of crystallisation temperatures and Turnbull parameters.*

Phys. Chem. Glasses: Eur. J. Glass Sci. Technol. B, 2025, **66**(4), 119 DOI: 10.13036/17533562.66.4.AM07

Structure of lithium alumino-titano-phosphate $Li_2O–Al_2O_3–TiO_2–P_2O_5$ (LATP) glasses as possible electrolytes for lithium batteries

Gavin Mountjoy,[1] *Marie Naylor,*[1,2] *Gregory Tricot,*[2] *Maria Alfredsson*[1] *& Sylvain Cristol*[2]

[1] *School of Engineering, Mathematics and Physics, University of Kent, Ingram Building, Canterbury CT27NH, UK*
[2] *LASIRE Unité Mixte de Recherche, Campus Cité Scientifique, Université de Lille, 59655 Villeneuve d'Ascq, France*

Corresponding author email: g.mountjoy@kent.ac.uk

Li ion technology is the first choice for future battery technology, and a number of advances are being sought. Changing from liquid to solid electrolytes[1] *improves safety and sustainability by removing organic volatiles. Introducing glass into the electrolyte lowers the working temperature which reduces energy expenditure and enables more applications. Lithium-based glass-ceramics have potential to be used as solid electrolytes.*

This presentation will be about lithium alumino-titano-phosphate (LATP) glasses with potential for Li ion conductivity.[2,3] *Interestingly, some LATP glasses have been reported with the analogous composition to "NASICON" materials for superionic conductivity of Na. On the other hand, knowledge of the LATP glasses is better assembled by understanding the effects of adding Al and/or Ti to the base lithium metaphosphate composition.*

This talk will present initial work done has been on glass synthesis of LP, LAP, LTP and LATP compositions. This has been accompanied by molecular dynamics modelling with rigid ion potentials, and infrared and Raman spectroscopy to identify Q^n units in the phosphate network. In addition, a comprehensive nuclear magnetic resonance (NMR) spectroscopy (1D and 2D, ^{27}Al and ^{31}P) is being used to get more details about the environment of network formers. This talk will present the background and initial results on LATP glass structures.

References

1. Zhai H. et al (2017) *Nano Letters* 17, 3182.
2. Leo C.J. (2002) *Materials Research Bulletin* 37, 1419.
3. Manthiram A. et al (2017) *Nature Review Materials* 2, 16103.

Phys. Chem. Glasses: Eur. J. Glass Sci. Technol. B, 2025, **66**(4), 120 DOI: 10.13036/17533562.66.4.AM08

Dissolution of phosphate glass examined through computer simulation using reactive force fields

Zohreh Fallah & Jamieson K. Christie*

Department of Materials, Loughborough University, Loughborough LE11 3TU, UK
**Current address: Department of Physics, Nanoscience Center, University of Jyväskylä, Jyväskylä, 40500, Finland*

Author emails: z.fallah@lboro.ac.uk, j.k.christie@lboro.ac.uk

Phosphate glasses have applications in optics, nuclear waste encapsulation, and biomedicine, depending on the composition and inclusion of therapeutic ions. The use of phosphate glass in many of these applications is controlled by their dissolution rate, which can vary by several orders of magnitude, depending on composition.

Computer simulation has been very helpful in complementing experimental studies of all sorts of glass structure and properties, including phosphate glass dissolution. In order to simulate the effects of composition and structure on dissolution correctly, we developed a reactive interatomic force field[1,2] (using the ReaxFF potential) by fitting against a very large set of training data, derived by quantum-mechanical DFT simulations.

We have shown that the potential performs well on simulations of the glass alone. Radial and angular distribution functions and coordination numbers of network formers and modifiers, as well as the network connectivity of the glass, are in agreement with experimental and previous simulations using both shell-model classical force fields and ab initio simulated data; for example, the coordination number of phosphorus is 4·0, and there is a distinct split between bonding to bridging and non-bridging oxygen atoms. However, the Q^n distributions are broader than expected.

We will discuss the effect of the accuracy of the potential on glass, as well as our ongoing work to use the potential to conduct simulations of the glass dissolving directly, that is, the interface of the glass with body fluid, represented in the first instance by water, and the likely future work in this area.

References

1. Fallah, Z. & Christie, J. K., *J. Chem. Phys.* **160**, 184505 (2024).
2. Fallah, Z. & Christie, J. K., *J. Phys. Chem. B* **128**, 11460 (2024).

Phys. Chem. Glasses: Eur. J. Glass Sci. Technol. B, 2025, **66**(4), 121 DOI: 10.13036/17533562.66.4.AM09

Combining machine learning, simulations and experiments for improved glass design

N. M. Anoop Krishnan[1]

[1] *Department of Civil Engineering, Indian Institute of Technology Delhi, Hauz Khas, New Delhi, India 110016*

Corresponding author email: krishnan@iitd.ac.in

The design of novel glasses with tailored properties remains a grand challenge in materials science, traditionally requiring extensive trial-and-error experimentation. This talk presents a comprehensive framework that integrates machine learning, atomistic simulations, and targeted experiments to accelerate the discovery and optimization of glass compositions. We demonstrate how large-scale machine learning models trained on experimental databases can predict glass properties across nearly the entire periodic table, enabling rapid screening of promising compositions. These predictions are refined through physics-informed graph neural networks that capture atomic-scale interactions and energy conservation principles, providing deeper insights into structure-property relationships. The framework is enhanced by automated knowledge extraction from scientific literature using our materials-specific language model (MatSciBERT) and natural language processing techniques. We showcase practical applications including the development of radiation-resistant nuclear waste glasses, scratch-resistant display glasses, and specialty glasses for space applications. The talk concludes by highlighting how this integrated approach has led to the discovery of novel glass compositions with superior properties, validated through experimental synthesis and characterization. Our results demonstrate that the synergistic combination of artificial intelligence, computational modeling, and experimental validation can significantly accelerate the glass design process while reducing resource requirements and environmental impact.

Phys. Chem. Glasses: Eur. J. Glass Sci. Technol. B, 2025, **66**(4), 122 DOI: 10.13036/17533562.66.4.AM10

Clustering of iron in glasses: composition–structure–property relations

Paul A. Bingham, Jack Cornwall, Amy Young, Alex Scrimshire & James D. Eales*

[1] *School of Engineeering and Built Environment, Sheffield Hallam University, City Campus, Howard Street, Sheffield S1 1WB, UK*

* *Corresponding author email: p.a.bingham@shu.ac.uk*

Ionic clustering phenomena in glasses have long been known and studied. However, the nature, extent, impact on properties and effects of bulk glass composition and structure have yet to be elucidated. Suzdalev & Imshennik[(1)] discussed how glassy matrices can incorporate transition metals distributed both heterogeneously in clusters and homogeneously throughout the glass matrix, and this has been shown previously using different spectroscopic techniques. Moon et al[(2)] defined clustering in glasses as more than one ion coupled by exchange interactions. Yet establishing the length scales associated with these exchange interactions, and the effects of the bulk glass itself and redox conditions, ion type and other factors, remain elusive. Here we have studied a range of glass types (silicate, borosilicate, borate and phosphate), doped with 0-5 mol% Fe_2O_3 and analysed using ^{57}Fe Mössbauer spectroscopy, X-band Electron Paramagnetic Resonance (EPR) spectroscopy, and other characterisation techniques, with the aim of understanding the onset, extent and relative proportions of clustered Fe ions in these glasses. The results to date will be presented and discussed.

References

1. I. P. Suzdalev and V. K. Imshennik, "Magnetic relaxation properties and clustering of iron ions in polymers and glasses," Nucl. Instruments Methods Phys. Res., 199 (1982) 125–132
2. D. Moon, J. Aitken, R. MacCrone, and G. Cieloszyk, "Magnetic properties and structure of XFe_2O_3, (1-X)($BaO.4B_2O_3$) glasses, Phys. Chem. Glasses 16 (1975) 91–102.

Phys. Chem. Glasses: Eur. J. Glass Sci. Technol. B, 2025, **66**(4), 123 DOI: 10.13036/17533562.66.4.AM11

Molecular dynamics simulations of the environment of alkali ions in mixed lithium-sodium silicates

Alastair N. Cormack

Inamori School of Engineering, NY State College of Ceramics, Alfred University, 1 Saxon Drive, Alfred, NY 14802 USA

Corresponding author email: cormack@alfred.edu

The mixed alkali effect, in which there is a non-linear change in the physical properties of, say, an alkali silicate glass when one alkali metal is replaced with another, is well known, having been widely observed experimentally.

Although there have been a number of attempts to provide a theoretical explanation, none has received universal acceptance. In large part, this is probably because the understanding of the structure of glasses at the time was limited to the ideas of e.g. Zacharaiasen, and Warren.

However, the last thirty years or so have seen significant advances in our understanding of the atomic structure of glasses, made possible by developments in, inter alia, x-ray and neutron scattering, solid state NMR and atomistic computer simulations.

We have used molecular dynamics to investigate the structure of several series of mixed lithium and sodium silicates, with the object of elucidating the detailed atomic scale environment of the alkali ions as a function of both Li/Na ratio and silica content. Di-, tri-, and tetra-silicates have been modelled, with the following Li/Na ratios: 100% Li, 95Li05Na, 75Li25Na, 50Li50Na, 25Li75Na, 05Li95Na and 100% Na.

In this presentation, we will discuss the changes in structure as a function of Li/Na ratio. Focus will be on the environments of the alkali ions.

Phys. Chem. Glasses: Eur. J. Glass Sci. Technol. B, 2025, **66**(4), 124 DOI: 10.13036/17533562.66.4.AM12

Johari–Goldstein relaxation in rejuvenated chalcogenide glasses

Jacopo Baglioni[1]* *& Giulio Monaco*[1]

[1] *Department of Physics and Astronomy "Galileo Galilei", University of Padova, Via F. Marzolo, 8 - 35131 Padova, Italy*

* *Correspondng author email: jacopo.baglioni@unipd.it*

When a liquid is cooled down to temperatures close to the glass transition, the relaxation dynamics are characterized by two timescales associated with the structural relaxation and a secondary process known as the Johari–Goldstein (JG) relaxation. The JG relaxation is related to many crucial properties of glasses, such as their plastic response, and is here investigated using fast-scanning calorimetry in rejuvenated, high-enthalpy $GeSe_3$ glasses. Rejuvenation is achieved by two methods: (i) increasing the cooling rate used to quench the melt, and (ii) irradiating the glass with x-rays. Both methods make the JG relaxation visible in the calorimetric traces, where it appears as an exothermic signal at temperatures below the glass transition. The JG relaxation can be linked to strained regions produced by quenching or to defected regions produced by irradiation. These findings strongly support a general connection between the JG relaxation and local defected regions in the glass network, and offer a strategy to control its strength.

Phys. Chem. Glasses: Eur. J. Glass Sci. Technol. B, 2025, **66**(4), 125 DOI: 10.13036/17533562.66.4.AM13

Bond angles and the structure of vitreous BeF_2

Alex C. Hannon[1]

[1] *ISIS Facility Rutherford Appleton Laboratory, Chilton, Didcot, Oxon OX11 0QX, UK*

Corresponding author email: alex.hannon@stfc.ac.uk

Vitreous BeF_2 has a structure formed of corner-sharing BeF_4 tetrahedra, which appears to be similar to that of vitreous SiO_2. The strength of Be–F bonds is much less than that of the Si–O bond, and thus BeF_2 may be considered a weakened analogue of SiO_2. BeF_2, however, is much less studied than SiO_2 due mainly to its high toxicity. A report is given of a neutron diffraction study of the structure of vitreous BeF_2, with particular emphasis on the bond angle distribution, both in BeF_2 and in other similar systems.

Phys. Chem. Glasses: Eur. J. Glass Sci. Technol. B, 2025, 66(4), 126 DOI: 10.13036/17533562.66.4.AM14

Where does sodium (Na) go in sodium silicate glasses? Implications for physical properties and behaviour

Grant S Henderson,[1] H.W. Nesbitt,[2] G.M Bancroft,[3] PAW Dean[3] & J Tse[4]

[1] Department of Earth Science, 22 Ursula Franklin Street, University of Toronto, Canada, M5S 3B1
[2] Department of Earth Sciences, University of Western Ontario, London, Ontario N6A 5B7
[3] Department of Chemistry, University of Western Ontario, London, Ontario N6A 5B7 [4]Department of Physics and Engineering Physics, University of Saskatchewan, Saskatoon, Saskatchewan S7N 5E2

Corresponding author email: henders@es.utoronto.ca

With a traditional modified random network approach describing the structure of alkali silicate glasses (Na-silicates), one assumes that Na_2O breaks bridging oxygens (BO) to form non-bridging oxygens (NBO) with the Na bonding to the NBO at the apex of a SiO_4 tetrahedron. In addition, the Na is then assumed to reside within the rings of SiO_4 tetrahedra making up the medium-range structure of the glass. However, this simplified approach is not consistent with more recent studies. We know that a network modifier can attach to the bridging oxygen(1) and, more than one modifier may bond to a single BO increasing the BO coordination from 2-3-4, etc. Our recent ab initio molecular orbital calculations are used to investigate the bonding behaviour of Na within the glass network. The results indicate that the modifier may also bond to the edges and faces of tetrahedra, and the coordination of a network modifier is composed of both BOs and NBOs from 1 or more adjacent tetrahedra. Furthermore, the Na resides within appropriately sized rings and within cage-like voids created by 3-dimensional ring structures such as those observed in crystalline cristobalite and stuffed tridymite structures. There are several implications for the physical behaviour of alkali-silicate glasses,(2) including partial molar volume and conductivity. The behaviour is a result of both inclusion of the modifier within rings and cages, as well as collapse of the modifier toward the surrounding oxygen atoms via electrostriction.

References

1. H.W. Nesbitt, G.M. Bancroft. G.S. Henderson, R.Ho, *JNCS*, **409,** 139 (2015).
2. H.W. Nesbitt, P. Richet, G.S. Henderson, G.M. Bancroft. *American J Science*, **325,** (2025)

Phys. Chem. Glasses: Eur. J. Glass Sci. Technol. B, 2025, **66**(4), 127 DOI: 10.13036/17533562.66.4.AM15

Structure and properties of high-density lead bismuth borosilicate glasses

B. Herrejon-Zavalza, Z. Pearlman & Mario Affatigato

[1] *Physics Department, Coe College, Cedar Rapids, IA 52402, USA*

Corresponding author email: maffatig@coe.edu

Lead bismuth borosilicate glass is a four-component glass system with the potential to reach very high densities at certain compositions. This research focuses on studying the physical properties and structure of these high-density samples. This work aims to find the maximum density possible from this glass family. The glasses followed the stoichiometry $x[yPbO.(1-y)Bi_2O_3].(1-x)[(1-z)SiO_2](1-x)(z)B_2O_3+(1-x)(z)3H_2O$. Experimentally, after heating the samples in a platinum crucible, the melts were cooled through plate-quenching and roller-quenching. For high-density samples, the temperature was held at 1000°C but went up to 1300°C for some lower-density samples. This led us to make glasses with x-values ranging from 0·1 to >0·9, y-values from 0·5< to 0·85, and z-values from 0 to 1. We also substituted aluminium oxide, tellurium oxide, and other glass formers to see the change in density and other properties. Density was measured using pycnometry and ranged from 3 g/cm^3 to 8·6 g/cm^3. We took refractive index measurements using a refractometer and, using a simultaneous thermal analyser, we measured the stability of glass in the samples. For high-density samples (density >8 g/cm^3), stability ranged from 15°C to 150°C. For details on structure, we completed Raman spectroscopy and Fourier transform infrared spectroscopy. We also carried out x-ray diffraction measurements on the roller-quenched samples to verify the absence of crystallinity. The bismuth oxide content was found to play an important role in enhancing the density, but it also lowered thermal stability. Incorporating greater amounts of silicon oxide alongside boron oxide played a similar role, growing the density at a thermal cost when the modifier fraction was kept constant.

This work was supported by the National Science Foundation under grants DMR-CER-2054930 and DMR-CER-2203142.

Phys. Chem. Glasses: Eur. J. Glass Sci. Technol. B, 2025, **66**(4), 128 DOI: 10.13036/17533562.66.4.AM16

Physical properties of sodium germanate glasses over an extended range of compositions

Zachary Pearlman,[1] Luci Green,[1,2] Rebecca Gabrielsson[1] & Steve A. Feller[1]

[1] *Physics Department, Coe College, Cedar Rapids, IA, 52402, U.S.*
[2] *Department of Chemistry and Biochemistry, University of South Carolina, Columbia, SC, 29208, U.S.*

Corresponding author email: sfeller@coe.edu

The sodium germanate glass family $[xNa_2CO_3+(1-x)GeO_2]\rightarrow[xNa_2O.(1-x)GeO_2+xCO_2]$ was explored up to x=0·7, where x is the molar fraction of sodium oxide in the glass. The results confirm previous research conducted by this laboratory(1) and emphasise the system at both ends of the glass forming region to understand complex changes in structure. To explore these changes, samples were made at x=0·05 intervals with smaller increments when x<0·1. The glass transition (T_g) and the onset of crystallisation (T_x) temperatures were measured using simultaneous thermal analysis (STA). Results from this study are consistent with previous work including the maximum in T_g around x=0·15, hypothesised to come from a coordination change in Ge and the anomalous T_g minimum below x=0·1, thought to be due to formation of NBOs bonded to Ge tetrahedra. Density measurements were taken on compositions with x<0·45 utilising helium pynometry. Packing fractions and free volumes were found for samples with density results. At high modifier contents above x=0·6, Raman spectroscopy was employed to measure CO_3 retention in the system by observation of the peak near 1080 cm^{-1}.

Acknowledgement

This work was supported by the National Science Foundation NSF DMR-RUI #2203142 and NSF PHY-REU #2349118.

Reference

1. T.J Kiczenski, Chris Ma, Eric Hammarsten, Doug Wilkerson, Mario Affatigato, Steve Feller, "A study of selected physical properties of alkali germanate glasses over wide ranges of composition," *J. Non-Cryst. Solids* **272** (2000) 57-66.

Phys. Chem. Glasses: Eur. J. Glass Sci. Technol. B, 2025, **66**(4), 129 DOI: 10.13036/17533562.66.4.AM17

Characterisation by ^{31}P NMR spectroscopy of acid glass of potassium phosphate

Fatma Zahra-Souissi,[1,2] *Maud Barré*[3] *& Toumi Mohamed*[1]

[1] *Solid State Physicochemical Laboratory (LES), University of Sfax, Faculty of Sciences of Sfax, Route de Sokra Km 3.5, BP 802, Sfax 3018, Tunisia*
[2] *Laboratory of Physics of Materials and Nanomaterials Applied at Environment, Faculty of Sciences, Gabes University, 6072, Gabes, Tunisia*
[3] *Oxides and Fluorides Laboratory (UMR 6010 CNRS), Molecular Engineering and Functional Materials Research Institute (FR CNRS 2575), University of Maine, Av. O. Messiaen, 72085 LE MANS Cedex 9, France*

Corresponding author email: ezzahramedenine@gmail.com

$2KH_2PO_4$–$1P_2O_5$ composite systems were prepared by a conventional solid-state method. XRD patterns of the synthesised composites revealed the formation amorphous character. The analysis of the internal stress in the composites systems by ^{31}P NMR indicates that the phases are 74% $K_1H_3P_2O_7$, 15% $K_2H_3P_3O_{10}$ and 11% $K_3H_3PO_4$.

Impedance spectra were well modelled. The obtained results show the electrical conductivity curve we have reported and that our sample exhibits two different behaviours. The observed semiconductor character has been attributed to the hopping motion of small polaron at high temperature and motion hopping of electron to variable ranges at low temperature.

Keywords: Potassium phosphate glass, ^{31}P NMR, XRD diffraction, DC elctrical conductivity

Phys. Chem. Glasses: Eur. J. Glass Sci. Technol. B, 2025, **66**(4), 130 DOI: 10.13036/17533562.66.4.AM18

From melt to glass stabilised below the glass transition temperature: structural changes

Natalia M. Vedishcheva & Nikolay S. Yuritsyn

Institute of Silicate Chemistry, Nab. Makarova 2, St. Petersburg, 199034, Russian

Corresponding author email: ionatali386@gmail.com

Changes in the structure in the transition from a stable melt to glass stabilized at temperatures below T_g are considered for the compositions $20Na_2O.80B_2O_3$ and $33{\cdot}3Li_2O.66{\cdot}7SiO_2$ whose T_g values are very close to 450°C. Structural changes are analysed over the intervals 1200–390°C for the borate composition and 1527–390°C for the silicate composition. The distribution of the basic structural units in both compositions and that of superstructural rings in the borate composition are calculated as a function of temperature using a thermodynamic approach based on the model of associated solutions, together with information about the structure of thermodynamically stable crystals existing in the borate and silicate systems. As opposed to the random network theory, this approach (known as the concept of the chemical structure) observes the requirement of the minimum Gibbs free energy of a given system. Due to this, the concept does not require the use of adjustable parameters. Since glass is a product of chemical interactions between its oxide components, structural changes in $20Na_2O.80B_2O_3$ and $33{\cdot}3Li_2O.66{\cdot}7SiO_2$ glasses with temperature are interpreted in terms of the different chemical nature of borate and silicate glass-forming systems.

The work was performed within the framework of the State Assignment on the project 1024030700034-2-1.4.3 (0081-2025-0003).

Phys. Chem. Glasses: Eur. J. Glass Sci. Technol. B, 2025, **66**(4), 131 DOI: 10.13036/17533562.66.4.AM19

Assessment of irreversibility in the performance of glass furnaces

R. Conradt

uniglassAC GmbH, Aachen, Germany

Corresponding author email: reinhard.conradt@uniglassac.de

Glass industry faces major challenges imposed by the concern for the global climate change. This calls for a reassessment of glass furnace concepts. Fundamentally, any thinkable furnace concept obeys the simple overall balance $P_{in}=P_{ex}+P_{loss}$. The symbol P denotes powers in units of kW; index "in" denotes the power input by fuel and electricity, index "ex" denotes the useful or exploited power determined by the intrinsic energy demand of melting and the heat content of the melt at exit temperature T_{ex}, index "loss" comprised the cumulative losses. Both P_{in} and P_{ex} depend on the pull rate p, with a maximum efficiency P_{ex}/P_{in} reached at a specific pull rate p_{opt}. The above power balance reflects both the first and second law of thermodynamics. In specific, P_{loss} corresponds to the "uncompensated heat" in Clausius' concept of entropy. It comprises both pull independent and pull dependent contributions. The former ones consist in the heat losses through the furnace periphery, while the latter ones are due to intrinsic entropy generation. The most important sources of intrinsic entropy generation are: the heat capacity flow imbalance between furnace atmosphere and melt, the resistance to heat transfer, and the chemical irreversibility of the batch-to-melt reactions. Optimal performance corresponds to minimum entropy generation. The issue is demonstrated with operational data of real furnaces.

Phys. Chem. Glasses: Eur. J. Glass Sci. Technol. B, 2025, **66**(4), 132 DOI: 10.13036/17533562.66.4.AM20

Unlocking glass sustainability

Steve Whettingsteel

Krysteline Technologies Ltd, Southampton, Hampshire, SO14 3JZ, UK

Corresponding author email: steve.w@krysteline.com

Over the past decade we have learnt that sustainability must become a substantial part of our global strategy, as resources dwindle and costs spiral legislation grapples with the most effective method of driving change without destroying a countries economy. Compared to plastic glass has been effective at reducing its production carbon footprint over the past seventy or more years, yet we struggle to nurture a global strategy to champion glass, often relying on product associations to create awareness, rather than a unified voice. The lack of a single, strong, and combative voice has allowed the unchallenged and dramatic adoption of plastic as the preferred packaging material. Understanding the complex challenges faced by glass and piecing together a strong and cohesive strategy which encompasses all glass types is vital in protecting the future of glass. Legislation often being developed with good intention, yet effective adoption often fraught with uneconomic or impractical constraints, compounded by the high cost of circularity being championed over the true cost of carbon reduction. Is circularity really possible, or should we be developing a clear perspective as to how an effective carbon assessment system could dramatically increase recycling and reuse of all glass types across multiple markets and territories. This is only possible with greater understanding and acceptance of the whole glass landscape from consumer electronics to PV Cells, packaging and construction, understand and influence how best we recover, process, and utilise the material, preventing its loss to landfill and generating sustainable long-term revenue is the cornerstone of a strong and stable resource recovery system. Academia must play a pivotal role in how the recovered resource could be used, partnering with industry to explore challenging areas and focusing on where the greatest impact can be achieved through strong, well-structured partnerships.

Phys. Chem. Glasses: Eur. J. Glass Sci. Technol. B, 2025, **66**(4), 133 DOI: 10.13036/17533562.66.4.AM21

Demand driven decarbonisation

Franck Crosnier

Senior Partner. edenseven, 5 High Green, Cambridge, CB22 5EG, UK

Corresponding author email: fcrosnier@edenseven.co.uk

This presentation will outline that Decarbonisation remains a critical agenda for all organisations. It will highlight that it is demand led and regulations will be aligned to net zero 2050 (e.g. MEES).

EPR evolution to include "modularity" illustrates how companies doing the right thing will be rewarded for their effort. Suppliers reducing their scope 1 and 2 are helping their clients scope 3, which can equate to approximately 80% of their total emissions.

It is critical to plan the pathway and have the software platform to measure and manage the process.

By having clear and accurate emissions data in place it will enable your business to reduce the **cost** *of operations, remain* **compliant** *with regulations and be* **credible** *to internal and external stakeholders.*

Done right, Decarbonisation programmes can transform your business and create significant benefits for your Customers.

Phys. Chem. Glasses: Eur. J. Glass Sci. Technol. B, 2025, **66**(4), 134 DOI: 10.13036/17533562.66.4.AM22

Powered by people: the path to net zero for glass is not just a technical challenge it is a human one!

Linda Newnes,[1] *Mey Goh*[2] *& Susan Lattanzio*[1]

[1] *Department of Mechanical Engineering, University of Bath, BA2 7AY, UK. Centre for People-Led Digitalisation (https://www.p-ld.co.uk/)*
[2] *Woolfson School of Mechanical, Electrical and Manufacturing Engineering, Loughborough University, LE11 3TU, UK. Centre for People-Led Digitalisation (https://www.p-ld.co.uk/)*

Corresponding author email: p-ld@bath.ac.uk

The route to a just transition to Net Zero is a challenge many organisations are facing – including UK glass manufacturing. The work by the team from British Glass (BG) has identified the policies and support required from the UK government to create the conditions to deliver a sustainable (economic and environment) glass manufacturing sector in the UK.

The Net Zero progress update by BG in 2024,(1) *demonstrates what is required to achieve the Glass sectors Net Zero ambitions, highlighting the policies and support required as well as the technical solutions needed. However, buy-in from the workforce and stakeholders is also an essential lever for success.*(2,3) *We need to ensure this 'change' is implemented on-site, in manufacturing plants in a way which is sustained and right first time. Why? Change takes time and we all know that people/stakeholders/the workforce can influence the level of success for any change programme. However, people and culture and getting this right is hard.*

In this talk Professor Newnes will introduce the importance of People-Led Change. Funded by the Industrial Challenge Strategy Fund, Innovate UK and UK Research and Innovation she will introduce why change is powered by people, providing examples of tools used for People-Led Digitalisation by companies such as Rolls Royce (Legitimacy of Proposed Change), Eyelit technologies (Manufacturing Execution Systems) and Domino Printing (Stakeholder Engagement). She will describe, through industry case studies, example tools which can be used to assist in deliver digital change and Net Zero within organisations.

Through these tools significant barriers to achieving sustainability in the glass sector may be addressed.

References

1. https://www.britglass.org.uk/sites/default/files/2024-12/British%20Glass%20-%20Net%20zero%20strategy%20update%20Nov%202024_0.pdf (accessed July 2025).
2. https://hbr.org/2025/06/its-time-to-update-how-your-company-talks-about-sustainability (accessed July 2025)
3. https://www.walkme.com/blog/change-management-in-manufacturing/#:~:text=Manufacturing%20is%20not%20a%20one,employee%20disengagement%20across%20all%20levels.

Phys. Chem. Glasses: Eur. J. Glass Sci. Technol. B, 2025, **66**(4), 135 DOI: 10.13036/17533562.66.4.AM23

Overview of CO_2 capture systems with hot potassium carbonate (HPC) applied to glass furnaces

Giorgio Di Lascio

K2-CO2 srl , Via DIEGO VALERI 23, Italia

Corresponding author email: info@k2-co2.com

Paper's topics: introduction of the CO_2 capture technical solution, comparison with alternative decarbonization techniques, maturity of the HPC capture process and its advantages, integration with the air pollution control system, advantages/disadvantages of its application to regenerative and oxyfuel furnaces, the example of a full-scale float plant and of a EU-LIFE funded project, quick overview of the way out for CO_2, typical CAPEX and OPEX incidence on glass production.

Phys. Chem. Glasses: Eur. J. Glass Sci. Technol. B, 2025, **66**(4), 136 DOI: 10.13036/17533562.66.4.AM24

Twenty years of experience in the valorization of glass pozzolan in concrete

Arezki Tagnit-Hamou

Industrial Chair on Glass Valorization in Materials, Civil Engineering and building Engineering Department, Sherbrooke University, Canada

Corresponding author email: a.tagnit@usherbrooke.ca

Portland cement is the most widely used construction material, but there are some limitations to using it for the production of concrete, i.e., its durability issues and emission of a high amount of CO2 in it production. So, in this study, the ground-glass pozzolan (GGP) is used to partially replace the cement for the in-situ application.

GGP obtained by grinding the mixed-waste glass to same fineness of cement can act as a supplementary-cementitious material (SCM), given that it is an amorphous and a pozzolanic material. The GGP showed promising performances in different concrete types such as conventional concrete (CC), high-performance concrete (HPC), and ultra-high-performance concrete (UHPC). The current paper reports on the characteristics and performance of GGP in different concrete types including as special field project of Darwin Bridges. These bridges won the 2021 Award of Excellence in the infrastructure category from the American Concrete Institute (ACI). The concrete is examined for the long term with respect to its performance by determining its fresh, hardened, visual and durability properties. It was found that concrete with the incorporation of GGP performs very well in all terms.

KEYWORDS: Ground Glass Pozzolan, Concrete, Supplementary Cementing Materials

Phys. Chem. Glasses: Eur. J. Glass Sci. Technol. B, 2025, **66**(4), 137 DOI: 10.13036/17533562.66.4.AM25

Adopting circularity in construction product manufacturing

Gareth Morris

Saint-Gobain UK & Ireland, Saint-Gobain House, East-Leake, Loughborough, LE12 6JU, UK

Corresponding author email: gareth.morris@saint-gobain.com

Moving to a circular economy is critical to preserving natural resources and supporting the pathway to net zero carbon, however it requires transformation in thinking, collaboration, and innovation across the value chain to be successful.

The construction industry is notoriously traditional and difficult to change but the last few years have seen an explosion in industry appetite for circular approaches, and there are a growing number of successful new business models facilitating the shift away from traditional approaches.

This presentation explores the role of a global construction product manufacturer in this shift, and discusses the opportunities, drivers, and challenges in adopting circular economy business models, with real life examples of success.

Phys. Chem. Glasses: Eur. J. Glass Sci. Technol. B, 2025, **66**(4), 138 DOI: 10.13036/17533562.66.4.AM26

Challenges and progress in recycling of specialty glass and glass-ceramics

Nils Eckl, Marius Amschler & Jörn Besinger

SCHOTT AG Mainz, Germany

Corresponding author email: joern.besinger@schott.com

SCHOTT AG is committed to UN's Sustainable Development Goals and focusses on SDG 3, 5, 12 and 13. Within the SDG 12 "Responsible Consumption and Production" we are running several pilot projects to close the loop for our products, such as secondary pharmaceutical packaging, recycling of pharma glass tubing and recycling of post-consumer glass-ceramic cooktop panels.

The presentation gives some insights into the general challenges along the value stream when it comes to the recycling of specialty glass and glass-ceramics. Further the pilot projects of pharma glass tubing and glass-ceramic cooktop panels are introduced, highlighting specific challenges resulting from the respective market and application requirements as well as from production technology restrictions.

In the glass-ceramic cooktop recycling project we are working on the reverse logistics to collect post-consumer glass-ceramic cooktops, disassembly and separation of the glass-ceramic, processing of the glass-ceramic to secondary raw material specification and reuse in our melting tanks to produce new glass-ceramic cooktop panels. The progress, current status and way ahead is described.

The pharma glass tubing project is a different case due to the fact that it is a highly regulated market. In the cause of this joint project with Sanofi, first time ever pre-industry material has been returned to our melting facilities and has been reused as cullet for new pharma tubing. The next step has already started with the goal to build up a closed loop for post-industry pharma glass packages, which is already contaminated with pharmaceuticals.

Phys. Chem. Glasses: Eur. J. Glass Sci. Technol. B, 2025, **66**(4), 139 DOI: 10.13036/17533562.66.4.AM27

Glass furnace lining recycling a challenge for sustainable refractory industry

Tiphaine Bazin[1] *& Michel Gaubil*[2]

[1] *VALOREF, ZI La Croisière, Bollene, France*
[2] *SEFPRO, SEPR, 2539 Route de Sorgues, Le Pontet FRANCE*

Corresponding author email: Tiphaine.Bazin@saint-gobain.com

Since decades VALOREF has developed solution to recover refractory lining from glass furnace. From dismantling glass plant, different technology for sorting refractories has been implemented and many possibilities define to reuse refractory material. Flow chart will be discussed during the presentation. This reclamation program concern many types of products such as dangerous and non-dangerous refractories, fibers, abrasives, dust from regenerators, blasting material and other ceramics.

By entrusting waste management to Valoref, glass industry can rely on solutions implemented to maximize the recycling of refractory waste while complying with the most stringent environmental standards.

Phys. Chem. Glasses: Eur. J. Glass Sci. Technol. B, 2025, **66**(4), 140 DOI: 10.13036/17533562.66.4.AM28

How the future of glass production could look

Stuart Hakes

F.I.C. (UK) Limited, Longrock Industrial Estate, Penzance, Cornwall, TR20 8HX UK

Corresponding author email: s.hakes@fic-uk.com

Much has been written and about in the media regarding decarbonisation, but what is the reality? The presentation will review the current options available for reducing CO_2 and examine their viability. The review will cover both furnace and downstream glass conditioning in all segments of the industry. It will take a realistic approach to the subject and offer a critical analysis of the steps taken so far by industry. The presentation will comment on how future developments ought to progress and discuss the barriers to be overcome.

Phys. Chem. Glasses: Eur. J. Glass Sci. Technol. B, 2025, **66**(4), 141 DOI: 10.13036/17533562.66.4.AM29

On the role of structural changes at the glass–liquid transition

Michael I. Ojovan[1] *& Dmitri V. Louzguine-Luzgin*[2]

[1] *School of Chemical, Materials and Biological Engineering, The University of Sheffield, Sheffield S1 3JD, UK*
[2] *Advanced Institute for Materials Research (WPI-AIMR), Tohoku University, Sendai 980-8577, Japan*

Corresponding author email: m.ojovan@sheffield.ac.uk

While the glass transition in amorphous materials is an obvious effect mechanically expressed by the solid-like behaviour such as brittleness of glasses against liquid-like behaviour including plasticity of molten state, the underlying atomic-size mechanisms and structural rearrangements responsible for the transition itself are still poorly understood. This resulted in widely spread claims that there is no structural difference between glasses and liquids.(1) *In the meantime the viscosity being even used on the logarithmic scale cannot serve as a criterion of glass transition because the viscosity at the calorimetric glass transition temperature region spans over four orders of magnitude from* $10^{8\cdot8}$ *to* 10^{13} *Pas at least.*(2) *Moreover, the treatment of vitrification as a process of continuously breaking ergodicity with entropy loss and a residual entropy tending to zero in the limit of zero absolute temperature disagrees with most experimental and theoretical investigations.*(3) *Structural rearrangements at calorimetric glass transition are behind drastic changes of materials characteristics causing differences of glasses from melts. Structural description of materials includes both species (atoms, molecules) and connecting bonds which are directly affected by changing conditions such as an increase of temperature when the bonding system becomes softer. At and above the glass transition a macroscopic percolation cluster made up of configurons (broken bonds) is formed which causes changes in the mechanical and thermal properties of the material from those of a solid to those of a liquid. An account of it enables the structural differentiation of glasses from melts. We have earlier emphasized the importance of distinguishing structural differences of amorphous materials below and above the glass transition temperature.*(4) *Here we highlight the role of the bonding system in understanding the nature of transformation of glasses into melts. Considering glass-liquid transition as melting of amorphous solids we formulate the melting criterion of solids as the condition of increase of Hausdorff-Besicovitch dimensionality of the set of configurons from zero to* $D_H = dim_H | S_B | \geq 2\cdot5 D_H = dim_H | S_B | \geq 2\cdot5$. *The constant volume jump of heat capacity at the glass transition is then shown to closely follow the equipartition theorem* $\Delta C_V \approx \Delta C_V^{trans} = 0\cdot5 D_H R$, *where R is the gas constant.*

References

1. J.D. Musgraves, J. Hu, L. Calvez (Eds.), *Springer Handbook of Glass*, Springer Handbooks, Springer Nature Switzerland AG 2019, Cham. 2019, 1851 p.
2. D.S. Sanditov, M.I. Ojovan, M.V. Darmaev. Glass transition criterion and plastic deformation of glass. *Physica B*, **582**, 411914, (2020).
3. J.W.P. Schmelzer, T.V. Tropin. Glass Transition, Crystallization of Glass-Forming Melts, and Entropy. *Entropy* **2018**, 20, 103
4. M.I. Ojovan, D.V. Louzguine-Luzgin. Revealing Structural Changes at Glass Transition via Radial Distribution Functions. *J. Phys. Chem. B*, **124** (15), 3186-3194 (2020).

Phys. Chem. Glasses: Eur. J. Glass Sci. Technol. B, 2025, **66**(4), 142 DOI: 10.13036/17533562.66.4.AM30

Analysis for boron using the "OXI" XRF standard analysis method at Sheffield Hallam University

Anthony M.T. Bell, Jack D. Cornwall, Miu Onaru, Casey A. Walsh & Paul A. Bingham

Materials and Engineering Research Institute, Sheffield Hallam University, Howard Street, Sheffield, S1 1WB, UK

Corresponding author email: Anthony.Bell@shu.ac.uk

The "OXI" XRF analysis method of Giles, Hurley and Webster(1), uses calibration standards to determine proportions of elements (present as oxides) in XRF fused bead (lithium borate glass) samples.

We recently re-established this method at Sheffield Hallam University using standard samples made using a Claisse LeNeo fused bead maker. In the past the lightest element that could be analysed by this technique was sodium. However, in 2022 Sheffield Hallam University obtained a Rigaku Primus IV XRF spectrometer, which can detect elements as light as boron. This presentation describes more borate glass standard samples were made and how "OXI" is being extended so that the lightest element that can be analysed by this technique is boron, to give accurate analyses of borate glass samples.

Reference

1. H.L. Giles, P.W. Hurley and H.W.M. Webster, X-ray Spectrometry, 24, 205-218 (1995).

Phys. Chem. Glasses: Eur. J. Glass Sci. Technol. B, 2025, **66**(4), 143 DOI: 10.13036/17533562.66.4.AM31

Vitrification of uranium tetrafluoride

Sarah Allinson,[1] *Nicholas Hodge,*[1] *Mike T. Harrison*[2] *& Dan Glymond*[2]

[1] *United Kingdom National Nuclear Laboratory, Preston Laboratory, Springfields, Salwick, Preston, Lancashire, PR4 0XJ, UK*
[2] *United Kingdom National Nuclear Laboratory, Central Laboratory, Sellafield, Seascale, Cumbria, CA20 1PG, UK*

Corresponding author email: mike.t.harrison@uknnl.com

The Sellafield Ltd (SL) Higher Active Waste Thermal Treatment Programme (HAWTTP) was set up to demonstrate the potential for vitrification as a cost-effective solution to complex intractable waste arisings that effectively destroys reactive components, reduces overall volumes, and results in a stable waste form as an end product. Following the completion of HAWTTP in 2024, a 3-year follow-on Thermal Treatment R&D programme was initiated by SL to continue to develop the technology and address specific knowledge gaps identified such as off-gas, glass composition and criticality concerns.

Whilst these SL programmes have focused on several arisings located at the Sellafield site, e.g. Magnox sludges, ion-exchange material (clinoptilolite) and plutonium contaminated waste (PCM), there are other materials in the UK's nuclear waste inventory that could also benefit from a thermal treatment process. For example, a number of drummed wastes currently stored at the Low-Level Waste Repository (LLWR) in Cumbria have been identified as having no final treatment route. These drums contain uranium tetrafluoride (UF_4) alongside other contaminants and cannot be disposed of in their current state. The suitability of processing options for these wastes at Springfields has been investigated; however, further process development is required.

Uranium generally has high solubility in silicate glass systems, especially in oxidising conditions. However, thermal treatment of UF_4 residues will release corrosive gases such as hydrogen fluoride (HF) or fluorine (F_2), which require complex and rugged off-gas treatment plant at full scale. Reducing or preventing the production of these off-gases at source simplifies the treatment process reducing plant operational costs. Ideally the fluoride gases would react with alkaline earth oxides, such as calcium and magnesium, and form stable compounds which would become immobilised within the vitreous phase.

This presentation details small scale experiments using complementary thermo-gravimetric analysis (TGA) and mass spectrometry (MS) gas analysis techniques to simulate encapsulation of UF_4 with the historic and new base glass compositions utilised for the vitrification of high-level waste (HLW) at Sellafield. The intention of these scoping experiments was to investigate how corrosive off-gases could be reduced via interaction with the glass matrix during the vitrification process.

Phys. Chem. Glasses: Eur. J. Glass Sci. Technol. B, 2025, **66**(4), 144 DOI: 10.13036/17533562.66.4.AM32

Behaviour and treatment of electrorefiner salt simulants during nuclear fuel processing

Lucas Greiner & Charmayne Lonergan

Missouri University of Science and Technology, Rolla MO 65409, USA

Corresponding author email: llgb4b@mst.edu

The safe immobilisation of electrorefiner salt waste (ERV), a byproduct of molten salt reactor (MSR) operations for spent nuclear fuel, is a critical challenge in the advancement of nuclear waste management technologies. ERV salts are highly enriched in alkali, alkaline earth, and rare earth fission product chlorides, necessitating dechlorination prior to stable waste form production. This study investigates the dechlorination and immobilisation of ERV salt simulants through thermally driven chloride-to-oxide conversion reactions under controlled atmospheric conditions. Thermodynamic modelling is employed to predict gas evolution behaviour, equilibrium stability fields, and the driving forces for halide volatilization as functions of temperature and atmosphere.

Characterisation of dechlorinated and immobilised products is conducted using various analytical techniques. high-performance liquid chromatography (HPLC) is utilized to understand phosphate structure evolution as the dechlorination/vitrification process proceeds. Mössbauer spectroscopy provides detailed insights into the oxidation state and coordination environments of iron species within the immobilised matrix. Raman and infrared (IR) spectroscopy characterise the evolution of molecular bonding, particularly the formation of the phosphate glass network. Differential scanning calorimetry (DSC) and thermogravimetric analysis (TGA) are used to track thermal transitions, mass loss behaviour, and correlate gas evolution events with dechlorination steps.

Preliminary results suggest that significant chloride removal is achieved within targeted thermal windows, 200-400°C, corresponding to specific volatility thresholds predicted by modelling. This integrated approach of thermodynamic prediction and experimental validation lays the groundwork for optimising ERV salt immobilisation processes, minimising halide retention, and ensuring the long-term chemical durability and stability of the resulting nuclear waste forms.

References

1. Murray, P. M.; Werth, C. J.; Riley, B. J.; Crum, J. V.; McCloy, J. S. *Phosphate-Based Dechlorination of Electrorefiner Salt Waste Using a Phosphoric Acid Precursor*. ACS Omega **2024**, *9* (12), 11234–11245. https://doi.org/10.1021/acsomega.4c00542.
2. Riley, B. J.; Pierce, D. A.; Chun, J.; Callow, R. A.; McCloy, J. S. *Development of Waste Forms for Electrochemical Salt Waste Containing Fission Products*. Ind. Eng. Chem. Res. **2020**, *59* (18), 8605–8613. https://doi.org/10.1021/acs.iecr.0c01357.

Phys. Chem. Glasses: Eur. J. Glass Sci. Technol. B, 2025, **66**(4), 145 DOI: 10.13036/17533562.66.4.AM33

The vitrification of waste containing metallic uranium from the nuclear industry

Jessica L. Blenkinsop,[1] Mike T. Harrison[1] & Helen Steele[2]

[1] United Kingdom National Nuclear Laboratory, Central Laboratory, Sellafield, Seascale, CA20 1PG, UK
[2] Sellafield Ltd, Sellafield, Seascale, CA20 1PG, UK

Corresponding author email: jessica.blenkinsop@uknnl.com

As part of the clean-up of legacy nuclear sites to ensure safe and responsible decommissioning, there will be a requirement to retrieve nuclear waste from ponds and silos. Waste will need to be processed into a stable and durable wasteform suitable for long-term disposal. Use of vitrification to form a durable glass product may be appropriate for some waste streams. However, some waste streams will likely contain unconfirmed quantities of metallic uranium that will require processing. A knowledge gap for the incorporation of uranium metal into a glass wasteform has been identified, which presents a risk for the acceptance of vitrification as a suitable mechanism for the immobilisation of various intermediate level waste (ILW) streams to aid decommissioning. A review has been carried out to identify how the nuclear industry could implement learning from other industries, national labs and academia, to better understand the incorporation of uranium into a glass wasteform. This work also explores the advantages of using glass as a host material for the complex waste streams present at nuclear facilities, and the research which would be required to allow vitrification to be accepted as the best available technology for uranium-containing streams. Essential work to close the knowledge gap would include, but would not be limited to, glass composition, melter technologies, durability testing of vitrified products, uranium redox chemistry, and limitations on the loading of uranium metal into melts.

Phys. Chem. Glasses: Eur. J. Glass Sci. Technol. B, 2025, **66**(4), 146 DOI: 10.13036/17533562.66.4.AM34

Femtosecond laser induced plasma modification of silicate glass

Gin Jose

School of Chemical and Process Engineering, University of Leeds, Leeds, LS2 9JT, UK

Corresponding author email: g.jose@leeds.ac.uk

Silicate glasses are by far the most used glass type for packaging, displays, architectural and specialty applications. Applications such displays and photonic components would require doping or surface engineering to create special functionalities such as strength enhancement and light emission/absorption. Use of lasers, in particular femtosecond (fs) lasers, for such applications is rapidly growing. It was the laser induced refractive index change that was widely explored for creating optical waveguides in glasses, which are either doped or undoped with rare earth ions, to fabricate passive or active functionalities. In the last decade years my research team has extensive studied a new technique called ultrafast laser plasma doping (ULPD) method to modify the surface of silica, silicate glasses and silica-on-silicon.[1] *The method is a powerful tool to introduce strength enhancement, color and refractive index change in silicate glasses. The ability to dope rare earth ions such as* Er^{3+} *in record high concentrations and to larger depths of several micrometers has opened the opportunity to engineer high gain per unit length erbium doped waveguide amplifiers for the C-Band of fibre optics communication.*[2–4] *The ULPD has been demonstrated with femtosecond laser of different pulse durations and repetition rates. These include fs-lasers at 1 kHz, 10 kHz and higher repetition rate lasers (upto 1 MHz) with pulse durations of 100 fs, 45 fs and <290 fs using different laser systems.* Er^{3+}*-ions doping has also been demonstrated in silica-on-silicon substrates as well. The highest doping in silica is demonstrated with high calculated gain per unit length and low losses to fabricate waveguide amplifiers.*[5] *The doping process essentially involve deposition of tellurite-based glasses with required dopant on a silica-based substrate at an elevated temperature below their glass transition temperature. In the talk an overview of the femtosecond laser induced changes in glass with emphasis on ULPD process will be presented.*

References

1. G. Jose, T. Fernandez, P. Steenson, and A. Jha, "Multi-ion diffusion in silica glass using femtosecond pulsed laser deposition," in Conference on Lasers and Electro-Optics 2012, OSA Technical Digest (online) (Optica Publishing Group, 2012), paper CM3L.6.
2. J. Chandrappan, M. Murray, T Kakkar. et al. "Target dependent femtosecond laser plasma implantation dynamics in enabling silica for high density erbium doping," Sci Rep 5, 14037 (2015).
3. S. Ahmad Kamil, J. Chandrappan, M. Murray, P. Steenson, T. F. Krauss, and G. Jose, "Ultrafast laser plasma doping of Er3+ ions in silica-on-silicon for optical waveguiding applications," Opt. Lett. 41, 4684-4687 (2016).
4. P. Pal, Eric Kumi-Barimah, B. Dawson, G. Jose, "Manufacturing of Er3+-doped planar waveguides on silica-on-silicon using femtosecond laser-induced plasma," Optics Comm., 522, 128614 (2022).
5. J. Chandrappan, M. Murray, P. Petrik, E. Agocs, Z. Zolnai, A. Tempez, S. Legendre, D. P. Steenson, A. Jha, and G. Jose, "Doping silica beyond limits with laser plasma for active photonic materials," Opt. Mater. Express 5, 2849-2861 (2015).

Phys. Chem. Glasses: Eur. J. Glass Sci. Technol. B, 2025, **66**(4), 147 DOI: 10.13036/17533562.66.4.AM35

The effect of impurities and irradiation on the glass network – Cs_2O-loaded iron phosphate case study

Cillian Cockrell

School of Computer Science and Engineering, Bangor University, Dean Street, Bangor, LL57 1UT, UK

Corresponding author email: c.cockrell@bangor.ac.uk

The use of iron phosphate glass as a wasteform is contingent both on its response to the addition of waste products and on its evolution in response to radiation emitted by these waste products. Both of these factors present significant alterations to the glass network and therefore our understanding of the resulting glassy state. We perform molecular dynamics simulations to study the effect of caesium, a nuclear decay product, on the glass network of a vitreous wasteform, both in pristine glass and after high-energy nuclear recoil cascades. Specifically, we simulate overlapping 70 keV cascades and examine the structural and topological effects that caesium has on the iron phosphate glasses before and after these cascades. We find that the glass network is substantially altered by the presence of caesium as a potent network modifier (though we hesitate to assign to it this label) and that radiation cascades produce qualitatively different effects from those in pure iron phosphate glasses. Overlapping cascades produce minimal effects on the mobility of caesium at low loading – radiation damage is much more severe in pure iron phosphate glasses. At higher loading, the glass network accommodates caesium atoms less well, particularly after irradiation. We explain this in terms of caesium's role as an "excluded" network modifier in comparison to iron, which is tightly incorporated into the glass network, as can be seen in Figure 1(a). The network-disrupting effect of the caesium anticipates the similar effect of radiation damage, causing the microscopic structure to resist alterations from radiation cascades. However, at large enough concentrations, the space demanded by caesium atoms may cause them to aggregate and be expelled from the glass phase entirely.

Figure 1(a): The larger caesium atoms do not fit tightly into the glass network, and terminate many phosphate chains. Iron atoms are less destructive, and fit tightly into the network

Phys. Chem. Glasses: Eur. J. Glass Sci. Technol. B, 2025, **66**(4), 148 DOI: 10.13036/17533562.66.4.AM36

On the thermal and kinetic methods of nepheline precipitation within Hanford high-level nuclear waste canisters

Jack D. Cornwall, Alex Scrimshire,[1] Anthony M.T. Bell,[1] Robin Smith,[1] Albert A. Kruger[2] & Paul A. Bingham[1]

[1] *School of Engineering and Built Environment, Sheffield Hallam University, Sheffield S1 1WB, UK*
[2] *Pacific Northwest National Laboratory, U.S. Department of Energy, Office of River Protection, 99352, Richland, WA, USA*

Corresponding author email: Jack.D.Cornwall@student.shu.ac.uk

Many nations plan to vitrify their nuclear waste in a borosilicate glass matrix for deep geological disposal.(1) Efforts to increase waste loading are crucial as cost savings due to the reduction of raw glass-forming chemicals and melt energy requirements, as well as for the reduction in geological repository volume that would need to be dug out to store the waste materials for their radioactive lifetimes.

Sodium and aluminium are two of the most plentiful elements present in wastes at Hanford, a nuclear waste site in southeastern Washington state, USA, due to their use in cladding and associated chemicals being used as part of the reprocessing effort.(2) When it comes to high-level waste (HLW), the limiting factor to waste loading is often that high alumina and high sodium oxide glass compositions are likely to precipitate nepheline during the slow canister centreline cooling (CCC).(3) Nepheline is a sodium aluminosilicate phase which tends to strip more network formers than modifiers from the adjacent glass,(4) which can result in said glass being more prone to water leaching if the steel containment canister fails.

Models to predict nepheline precipitation, such as the nepheline discriminator,(5) optical basicity,(6) the submixture model,(7) and the difference based on correlation(8) models, have gone some way to allowing higher waste loading of HLW glasses, however, they are often found to be overly conservative. This research looks at the thermal and kinetic evolution of crystals during CCC cooling along the compositional knife edge predicted by these models and attempts to explain why the models are conservative and how to address this.

References

1. Thorpe, C. L. *et al.* Forty years of durability assessment of nuclear waste glass by standard methods. *npj Mater. Degrad.* **5**, (2021).
2. Goel, A., McCloy, J. S., Pokorny, R. & Kruger, A. A. Challenges with vitrification of Hanford High-Level Waste (HLW) to borosilicate glass – An overview. *J. Non-Crystalline Solids X* **4**, 100033 (2019).
3. Petkus, L. CANISTER CENTERLINE COOLING DATA, REVISION 1. 1–3 at (2003).
4. Li, H., Vienna, J. D., Hrma, P., Smith, D. E. & Schweiger, M. J. Nepheline Precipitation in High-Level Waste Glasses : Compositional Effects and Impact on the Waste Form Acceptability. *MRS Proc.* **465**, 261 (1996).
5. Li, H. *et al.* Effects of Al2O3, B2O3, Na 2O, and SiO2 on nepheline formation in borosilicate glasses: Chemical and physical correlations. *J. Non. Cryst. Solids* **331**, 202–216 (2003).
6. Mccloy, J. S., Schweiger, M. J., Rodriguez, C. P. & Vienna, J. D. Nepheline Crystallization in Nuclear Waste Glasses: Progress Toward Acceptance of High-Alumina Formulations. *Int. J. Appl. Glas. Sci.* **2**, 201–214 (2011).
7. Vienna, J. D., Kroll, J. O., Hrma, P. R., Lang, J. B. & Crum, J. V. Submixture model to predict nepheline precipitation in waste glasses. *Int. J. Appl. Glas. Sci.* **8**, 143–157 (2017).
8. Sargin, I., Lonergan, C. E., Vienna, J. D., McCloy, J. S. & Beckman, S. P. A data-driven approach for predicting nepheline crystallization in high-level waste glasses. *J. Am. Ceram. Soc.* **103**, 4913–4924 (2020).

Phys. Chem. Glasses: Eur. J. Glass Sci. Technol. B, 2025, **66**(4), 149 DOI: 10.13036/17533562.66.4.AM37

Thermal treatment of radiological asbestos-containing wastes (ACW): a disposability toolbox

Hannah E. Blystra, Anthony M. T. Bell, Wei Deng & Paul A. Bingham

Materials and Engineering Research Institute, Sheffield Hallam University, City Campus, Howard Street, Sheffield S1 1WB, UK

Corresponding author email: hannah.blystra@student.shu.ac.uk

This project contributes to improvements in the following three key categories. One, current issues in the retrieval, sorting, segregation, and volumes of the Very Low Level Wastes (VLLW), Low Level Wastes (LLW) and Intermediate Level Wastes (ILW) further contaminated by asbestos.[1] Two, address the scarcity of available research on the management of radiological asbestos-containing wastes (Scopus shows <15 articles have been published between 1970 and 2022 on this type of asbestos treatment). Finally, three, assist in evaluating alternative lower CO_2 waste packaging. Vitrification of contaminated asbestos containing radioactive wastes seems a promising answer to all three of these categories, as vitrification creates a leach resistant product while lowering the volume of waste and immobilising radionuclides efficiently.[2,3]

This study began with an investigation into suitable chemical and morphological asbestos simulants for safer laboratory-based experiments. Then, several glasses were selected for compatible chemical composition and representation of current nuclear waste vitrification packages. These glasses were used in frit form, waste loaded at varying percentages, and heat treated to build a range of known workable packaging options. Once the waste loaded glasses were formed, they were analysed using XRF, XRD, Raman, and DTA in order to understand their advanced composition-structure-property relationship. Additional bespoke batch packages were also considered as a comparison for waste loading ability.

By integrating material science, process engineering, and regulatory frameworks, this study will aid in addressing critical waste management challenges in the nuclear sector. Contributing to a safer, more sustainable approach to ACW management, aligning with industry objectives and enhancing long-term environmental stewardship.

References

1. NDA, BEIS; *2022 UK Radioactive Waste Inventory*. 2022. Cumbria. UK Radioactive Waste and Material Inventory 2022 - GOV.UK [accessed 03.10.2023].
2. Ojovan, M. I., & Lee, W. E.; *Chapter 17: Immobilisation of Radioactive Waste in Glass. In An Introduction to Nuclear Waste Immobilisation.* 2014; pp. 245–282. DOI: 10.1016/B978-0-08-099392-8.00017-6
3. Fuks, L., Herdzik-koniecko, I., Kiegiel, K., Miskiewicz, A., & Zakrzewska-koltuniewicz, G.; *Methods of Thermal Treatment of Radioactive Waste*. Energies, 2022; Vol. 15, Issue 1. https://www.researchgate.net/publication/357628157_Methods_of_Thermal_Treatment_of_Radioactive_Waste. [accessed 03.10.2023].

Phys. Chem. Glasses: Eur. J. Glass Sci. Technol. B, 2025, **66**(4), 150 DOI: 10.13036/17533562.66.4.AM38

Rare earth optical materials

Setsuhisa Tanabe

School of Global Environment, Kyoto University, Kyoto, Japan

Corresponding author email: tanabe.setsuhisa.4v@kyoto-u.ac.jp

Since 1989, just before and after becoming an assistant professor in Kyoto U, most of my research was devoted to rare-earth doped luminescent materials such as upconversion lasers, optical amplifiers for fiber telecommunication, persistent/PSL phosphors, LED phosphors, thanks to the collaboration and meeting with many stimulating supervisors, colleagues and students. The material forms range from glass, ceramics and recently to transparent ceramics, that I love. The 2p abstract is too short to mention even part of my achievements and exciting experiences but I will try to present in 40 minutes talk on the day of this meeting of RESJ.

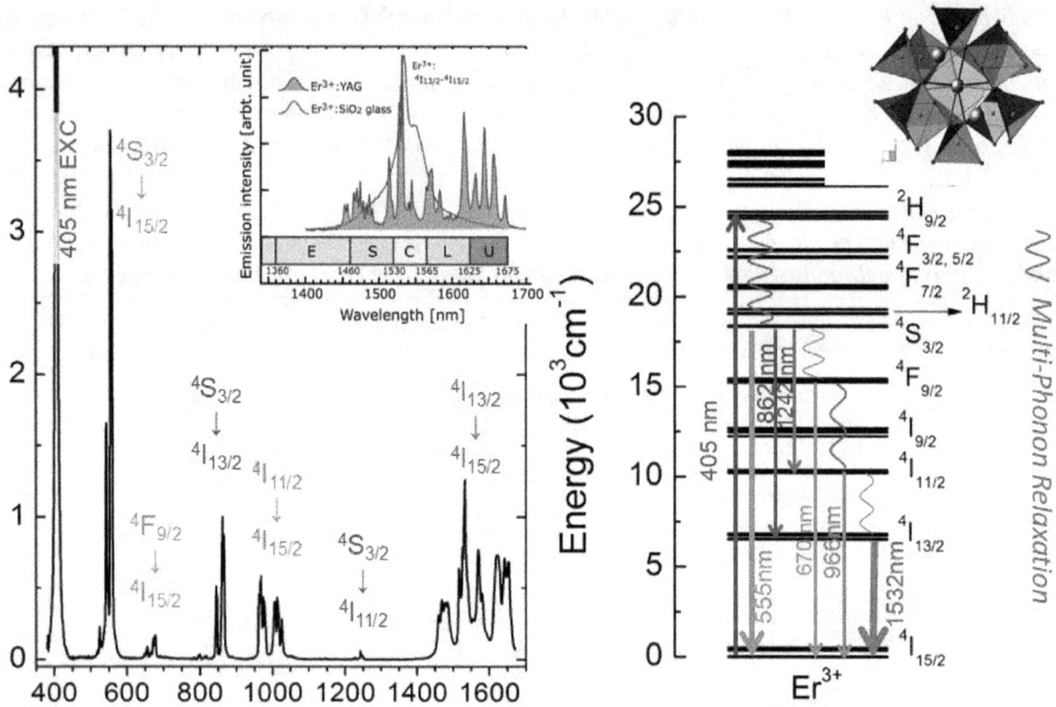

References

1. Tanabe, S. *Rare earths*, **41**, 1-10 (2006).
2. Ibid. **23**, 67-94 (1993). :
3. Ibid., **59**, 11-18 (2011).
4. Tanabe, S. *Int. J. Appl. Glass Sci.*, 000 (2015).
5. Xu, J. & Tanabe, S. *J. Lumin.* **205**, 581-620 (2019).
6. J. Ueda, K. Kuroishi, S. Tanabe, *Appl. Phys. Lett.* (2014).
7. Tanabe, S. *Applied Physics*, 79, 54-58 (2010)

Phys. Chem. Glasses: Eur. J. Glass Sci. Technol. B, 2025, **66**(4), 151 DOI: 10.13036/17533562.66.4.AM39

Integration of rare earth activated phosphor into inorganic glass: from bulk phosphor in glass to phosphor-glass thin films

Jeena Rose Jose,[1,2] *P. R. Biju*[2] *& Gin Jose*[1]

[1] *School of Chemical and Process Engineering, University of Leeds, Leeds, LS2 9JT UK*
[2] *Society School of Pure and Applied Physics, Mahatma Gandhi University, Kottayam, 686520, India*

Corresponding author email: jeenavattukattil@gmail.com

Light-emitting diodes (LEDs) have transformed the lighting industry owing to their energy efficiency, long lifespan, and design flexibility. However, the performance of commercially available LEDs is often hindered by the use of organic binders that encapsulate phosphor powders. These binders tend to degrade over time, resulting in reduced luminous efficacy and colour shifts. To address these limitations, phosphor-in-glass (PiG) composites have emerged as a promising alternative. PiGs are typically fabricated by mixing phosphor powders with glass frit and sintering the mixture at an optimal temperature to form a stable, inorganic matrix. Following the successful incorporation of YAG:Ce^{3+} into a glass matrix by Y.K. Lee et al. [1], a wide range of phosphors have since been investigated for PiG fabrication [2,3].

Building on this foundation, we explored the incorporation of luminescent phosphors into inorganic glasses based on borosilicate and tellurite, selected for their excellent chemical durability and thermal stability[4]. We adopted a low-temperature melt-quenching technique to embed the phosphors into the glass matrix. Tellurite-based glasses, in particular, proved to be highly suitable hosts for the phosphors due to their relatively low melting temperatures—which help minimize phosphor degradation—and their favourable refractive index compatibility with phosphors. The structural and microscopic characteristics of the resulting PiG composites were thoroughly examined using electron microscopy, supported by elemental mapping, confirming successful incorporation of phosphors within the glass matrix.

Despite successful synthesis and promising stability, challenges such as non-uniform phosphor distribution and a limitation in phosphor content (up to 20 wt%) affect the performance of bulk PiG materials. To address these drawbacks, we developed a novel thin-film "sandwich" architecture, embedding phosphor layers between a silica substrate and a tellurite glass coating achieved by the femtosecond pulsed laser deposition technique. This method not only ensures better diffusion and encapsulation of phosphor films but also significantly enhances luminescence efficiency. Our findings highlight the potential of these phosphor-glass films as robust, high-efficiency inorganic colour converters for next-generation lighting and display applications.

References

1. Y.K. Lee, J.S. Lee, J. Heo, W.B. Im, W.J. Chung, "Phosphor in glasses with Pb-free silicate glass powders as robust color-converting materials for white LED applications", Opt. Lett. 37, 3276 (2012).
2. J.S. Zhong, H.B. Gao, Y.J. Yuan, L.F. Chen, D.Q. Chen, Z.G. Ji, "Eu3+-doped double perovskite-based phosphor-in-glass color converter for high-power warm w-LEDs", J. Alloys Compd. 735,2303–2310 (2018).
3. P. Zeng, Z. Cao, Y. Chen, M. Yin, "Investigation of the temperature characteristic in SrB4O7:Sm3+ phosphor-in-glass by analyzing the lifetime of 684 nm, J. Rare Earths 35,783–786 (2017).
4. J.R. Jose, A. Gopinath, S.A. Oommen, C. Joseph, P.R. Biju, "Novel red emitting Phosphor-in-Glass (PiG) based on Na6CaP2O9:Eu3+ phosphor as an inorganic colour converter for w-LED applications", Opt. Mater. 148,114809 (2024)

Phys. Chem. Glasses: Eur. J. Glass Sci. Technol. B, 2025, **66**(4), 152 DOI: 10.13036/17533562.66.4.AM40

Tailoring structural and functional properties of phosphate glasses through CaO and Al_2O_3 incorporation

Yasmina Alaoui & Asmae Er-rafai

Laboratory of Materials, Nanotechnology, and Environment, Faculty of Sciences, Mohammed V University in Rabat, P.O. Box 1014, Agdal-Rabat, Morocco

Corresponding author email: alaouiyasmina18@gmail.com

Phosphate-based glasses are emerging as promising alternatives to silicate systems due to their compositional flexibility, lower processing temperatures, and potential applications in optics and durable matrices. This study synthesises two series of phosphate glasses via melt-quenching: $50P_2O_5–xCaO–(50−x)K_2O$ (x=0–40 mol%) and $xAl_2O_3–(50−x)P_2O_5–10CaO–40K_2O$ (x=2·5–15 mol%), to investigate the effects of CaO and Al_2O_3 on structural, thermal and optical properties.

X-ray diffraction confirmed the amorphous nature of all samples. FT-IR and Raman spectroscopy indicated a transition from P–O–P to more stable P–O–Ca and P–O–Al bonds, enhancing network connectivity. These structural changes corresponded with increased density and reduced molar volume. Differential scanning calorimetry revealed that higher Al_2O_3 content elevated the glass transition temperature and thermal stability, with no crystallisation observed. UV–Vis–NIR spectroscopy showed that Al_2O_3 variations allowed tuning of optical band gaps. Chemical durability tests demonstrated improved resistance to dissolution with increased Al_2O_3, especially in aggressive environments.

These findings suggest that potassium–calcium–alumina phosphate glasses offer a tunable platform for developing durable materials suitable for optical and environmental applications.

Reference

1. Alaoui, Y., Laourayed, M., Er-rafai, A., Hammi, M., El Moudane, M., Boudalia, M., Sekkat, Z., Warad, I., Guenbour, A., Bellaouchou, A., & Zarrouk, A. Effect of alumina insertion on structural properties, thermal stability, and chemical durability of potassium calcium based-phosphate glasses. *Inorganic Chemistry Communications*, 2022.

Phys. Chem. Glasses: Eur. J. Glass Sci. Technol. B, 2025, **66**(4), 153 DOI: 10.13036/17533562.66.4.AM41

Photo-induced PL colour change and anion migration of $CsPb(Br,I)_3$ perovskite nanocrystals in glass-ceramics

Kejie Huang,[1] *Ruilin Zheng*,[2] *Jian Xu*[1,3] *& Setsuhisa Tanabe*[1,3]

[1] *Graduate School of Human and Environmental Studies, Kyoto University, Kyoto 606-8316, Japan*
[2] *School of Science, Nanjing University of Posts and Telecommunications, Nanjing 210023, China*
[3] *Graduate School of Global Environmental Studies, Kyoto University, Kyoto 606-8316, Japan*

Corresponding author email: huang.kejie.74w@st.kyoto-u.ac.jp

Mixed halide perovskites have attracted significant attention due to their potential in solar energy applications. However, owing to their inherent structural instability, these materials may exhibit fluorescence color changes when exposed to light through phase segregation and lattice distortions[1,2] which can lead to degradation in device performance. This phenomenon has been studied both in thin film and single crystals, but very limited in glass-ceramics. We report a glass-ceramics containing $CsPb(Br,I)_3$ perovskite nanocrystals, which exhibits a fluorescence color change from yellow to red upon UV illumination (Figure 1(a). This transition suggests that the luminescent component undergoes a transformation from a bromide-rich phase to an iodide-rich phase. In a dark environment, the sample gradually recovers to pre-illumination state. The changes in fluorescence intensity under continuous illumination suggest that ion exchange and defect migration occur between the nanocrystals and the surrounding amorphous matrix (Figure 1(b).

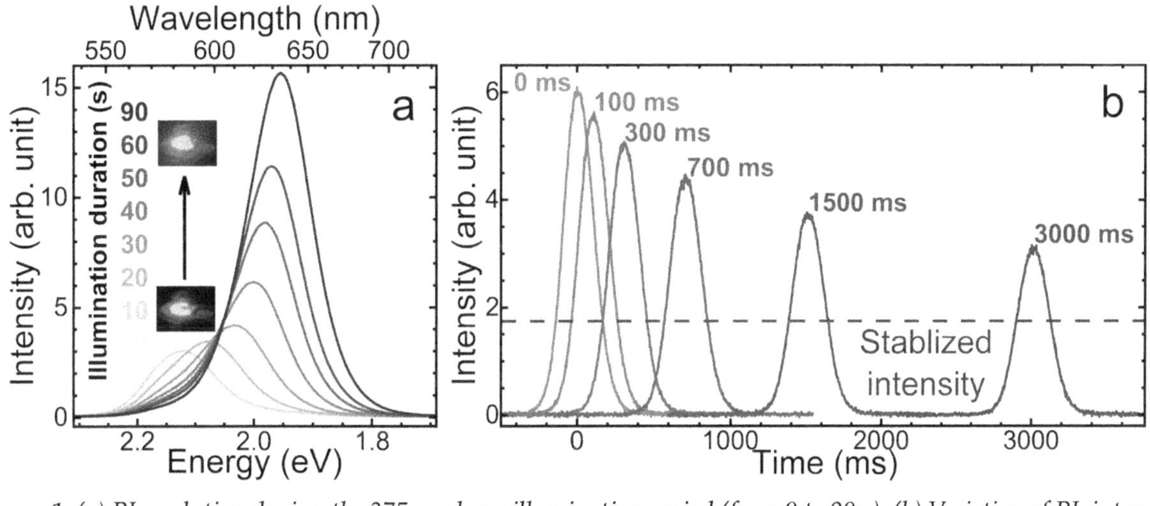

Figure 1. (a) PL evolution during the 375 nm laser illumination period (from 0 to 90 s); (b) Variation of PL intensity of the illuminated sample upon re-illumination after it was kept in a dark environment for 5 minutes

References

1. Hoke, E. T., Slotcavage, D. J., Dohner, E. R., Bowring, A. R., Karunadasa, H. I., & McGehee, M. D.. *Chemical Science*, 2015, **6**(1), 613-617.
2. Tominaka, S., Karimata, I., Matsuoka, T., Sakamoto, M., Nakajima, T., Ohara, K., & Tachikawa, T.. *ACS Energy Letters*, 2021, 6(11), 3858-3863.

Phys. Chem. Glasses: Eur. J. Glass Sci. Technol. B, 2025, **66**(4), 154 DOI: 10.13036/17533562.66.4.AM42

A journey from soft glasses to photonics devices

Yasutake Ohishi

Graduate School of Engineering, Toyota Technological Institute, 2-12-1 Hisakata, Tempaku, Nagoya 468-8511, Japan

Corresponding author email: ohishi@toyota-ti.ac.jp

Photonics devices, especially optical fibre devices, made of glasses have been developed for laser technology, optical telecommunications, optical measurement and optical sensing, optical quantum computing, and optical quantum information communication and processing, etc. Silica glass has been and are currently used as a major waveguide material for optical fibre devices. Many efforts have been being devoted to silica glass-based photonics device development, such as, optical fibres, waveguide devices, fibre lasers and amplifiers. On the other hand, the use of non-silica glass is expected to dramatically improve the performance of those photonics devices and may enable the realisation of properties that cannot be achieved with silica glass-based photonics devices.

This situation has led to the acceleration of research focused on new photonics devices using non-silica glasses. Among non-silica glasses, heavy metal oxide and non-oxide glasses have attracted particular attention. Fluoride[(1)] and chalcogenide[(2)] glass as non-oxide glass, and tellurite glass[(3)] as heavy metal oxide glass, which are called soft glasses, have especially attracted much attention as infrared transmission optical fibre materials, fibre laser hosts and nonlinear fibre materials.

Optical fibre device research using soft glasses started in earnest in the 1980s. We started this research from the exploration of glass compositions which are thermally stable and can be used for optical fibre glass. Next, we proceeded to develop fibre fabrication technology using soft glasses. This is because silica fibre technology could not be used for soft glass optical fibrerfabrication. We needed unique technology for soft glasses. We have demonstrated many prominent performances which cannot be realised by silica glass optical fibre devices using soft glass optical fibres. Representative examples are infrared transmission, rare-earth-doped optical fibre amplification and laser, Raman and Brillouin amplification, slow light generation, optical parametric amplification and supercontinuum generation, etc.

I have engaged in soft glass search, fibre structure design and fabrication, and optical performance measurement and analysis of optical fibre devices in the soft glass optical fibre device research. In this talk, I will present the major achievements of research on soft glass optical fibre devices to date and future prospect. I'd like to appreciate if you could feel a research journey in this field.

References

1. M. Poulain, M. Poulain, J. Lucas, and P. Brun, Mater. Res. Bull., 10(4), 243 (1975)
2. R. Frerichs, J. Opt. Soc. Am., 43(12), 1153, (1953) and R. Frerichs, Phys. Rev., 78 643 (1950).
3. J. E. Stanworth, Nature,169, 581 (1952).

Phys. Chem. Glasses: Eur. J. Glass Sci. Technol. B, 2025, **66**(4), 155 DOI: 10.13036/17533562.66.4.AM43

Development of a structural model of strontium doped bio-phosphates for the interrogation of medium-range order

Daniel Bradley,[1] *Emma Barney,*[1] *Emelie Nilsson,*[2] *Uresha Patel,*[1] *Ifty Ahmed*[1] *& Tristan Youngs*[3]

[1] *Advanced Materials Research Group, Faculty of Engineering, University of Nottingham, Nottingham, UK*
[2] *Department of Biomedical Science, Malmo University, Sweden*
[3] *ISIS Neutron and Muon Source, Harwell Campus, Didcot, UK*

Corresponding author email: emma.barney@nottingham.ac.uk

Phosphate-based glasses have received a lot of attention over the last two decades for their desirable qualities for biomedical applications. The biocompatibility and full resorbability of phosphate glasses make them suitable for use in hard and soft tissue engineering, bone fracture fixation, dental repair and controlled delivery of therapeutic ions. Current research has focused on addressing the specific clinical needs of different applications through compositional development. Multicomponent modifier-rich phosphate glasses are of specific interest due to their excellent biocompatibility and ability to be tailored for specific applications. Previous studies have used classic structural analysis such as NMR, scattering techniques, structural modelling studies, and functional property tests such as dissolution tests and cytotoxicity to understand the process-structure-property relationship, particularly focusing on the dissolution rates of the therapeutic ions.[(1)] *Many of the properties of bioactive phosphate glasses exhibit nonlinear behaviour as multiple different modifiers are substituted into the network. These mixed-modifier effects are thought to due to the medium range ordering of the modifier cations; something which is difficult to elucidate by directly fitting structural data.*

In this study a series of multi-component phosphate based glasses ($40P_2O_5.24MgO.20Na_2O.(16-x).CaO.xSrO$, where x=0, 4, 8, 12, 16) were measured using neutron diffraction. The structures of these glasses were simulated using empirical potential structure refinement (EPSR), implemented through the Dissolve software.[(2)] *To simulate such complex systems a new technique that refines a single of potentials across the entire compositional range of glasses has been developed to better constrain the structural models. This approach has enabled a detailed look at modifier environment, including modifier bonding and modifier–modifier clustering. Through this approach a more complete understanding of the mixed modifier effect can be gained, giving an insight into its influence on the physical properties of the system.*

References

1. Patel, U.; Moss, R. M.; Hossain, K. M. Z.; Kennedy, A. R.; Barney, E. R.; Ahmed, I.; Hannon, A. C. *Acta Biomaterialia* 2017, 60, 109-127
2. Youngs, T. *Molecular Physics* 2019, 117 (22), 3464-3477.

Phys. Chem. Glasses: Eur. J. Glass Sci. Technol. B, 2025, **66**(4), 156 DOI: 10.13036/17533562.66.4.AM44

Rhenium volatilization from molten glass: relationship between volatilization and viscosity

Chenchen Niu, Michael I. Ojovan & Kai Xu

State Key Laboratory of Silicate Materials for Architectures, Wuhan University of Technology, Wuhan 430070, P. R. China

Corresponding author email: kaixu@whut.edu.cn

Rhenium is used as a non-radioactive surrogate of technetium-99 (Tc-99), which is a highly volatile nuclide at elevated temperatures, posing critical challenges for nuclear waste vitrification. Understanding its volatilization behaviour is crucial for improving vitrification efficiency. This study investigates Re volatilization in borosilicate glass compositions with controlled viscosity. Reducing melt viscosity from 15·9 to 5·2 Pas increased the loss of Re by 40% over 8 h of melting time at 1200°C. In order to model the non-oxidizing volatilization, we developed a diffusion-reaction synergistical control model (NOX model) integrating Fick's law, reaction kinetics, and partition coefficients that characterizes the relationship between reaction and diffusion and accounts for unique localized states of Re in the glass melt. The NOX model effectively describes Re volatilization from molten glass across a wide range of melt viscosity. Parameter analysis shows that Re volatilization is governed by its diffusivity, and that Re diffusion coefficients increased from $1{\cdot}51\times10^{-8}$ *to* $1{\cdot}52\times10^{-7}$ cm^2/s *with decreasing melt viscosity. By modifying the Stokes-Einstein equation using a diffusion-equivalent radius of Re, we have revealed that:*

(1) Theoretical ionic radius of Re (~$6{\cdot}7\times10^{-11}$ *m) exceeds its diffusion-effective radius (~*$2{\cdot}1\times10^{-11}$ *m);*

(2) Lower viscosity reduces the Re diffusion-effective radius by 70%, indicating that the glass network depolymerization enhances Re diffusivity.

Meanwhile, rising partition coefficients under decreasing viscosity confirms that shorter diffusion pathways facilitate Re volatilisation in low-viscosity melts.

Phys. Chem. Glasses: Eur. J. Glass Sci. Technol. B, 2025, **66**(4), 157 DOI: 10.13036/17533562.66.4.AM45

Switching and thermal behaviour of Cu and Sn doped Si–Te glasses: a comparative analysis

B. Harini,[1] Diptoshi Roy,[2] K. B. Jagannatha,[3] S. Asokan[4] & Chandasree Das[1]

[1] *Dept. of Electrical and Electronics, BMS College of Engineering, Bangalore-560019, India and Visvesvaraya Technological University, Belagavi-590018, India*
[2] *Dept. of Electrical and Electronics Engg., CMR Institute of Technology, Bangalore-560037, India and Visvesvaraya Technological University, Belagavi-590018, India*
[3] *Dept. of Electronics and Communication Engg., BMS Institute of Technology and Management, Bangalore-560064, India*
[4] *Dept. of Instrumentation and Applied Physics, Indian Institute of Science, Bangalore-560012, India*

Corresponding author email: chandasreedas.eee@bmsce.ac.in

A comparison of switching and thermal behaviour of Cu (transition metal) doped and Sn (metal) doped Si–Te glasses are accomplished in this article. Conventional melt quenching technique is used for the synthesis of $Si_{15}Te_{85-x}Cu_x$ ($1\leq x\leq 10$) and $Si_{20}Te_{80-x}Sn_x$ ($1\leq x\leq 7$) bulk glasses. The amorphous nature of the bulk glasses is confirmed by x-ray diffraction whereas the compositional analysis is performed by EDAX.

Electrical switching behaviour of $Si_{15}Te_{85-x}Cu_x$ ($1\leq x\leq 10$) is connected to the dopant resistivity and network connectivity, with notable transitions from amorphous to crystalline phases during switching, impacting electrical properties.[(1)] $Si_{20}Te_{80-x}Sn_x$ ($1\leq x\leq 7$) samples display memory switching with the currents in the order of ≥4 mA and threshold switching with <4 mA of current.[(2)] Cu doping results in retention of memory switching whereas Sn doping exhibits both threshold switching and memory switching. The amorphous to crystalline phase transition during switching, occurs in both the glasses.[(1,2)]

Alternating differential scanning calorimetry (ADSC) is used for thermal analysis. Thermal studies on Cu–Si–Te glasses show the compositional dependence of non-reversing enthalpy at glass transition (HNR), revealing the existence of Boolchand intermediate phase (IP) in $2\leq x\leq 6$ range.[(3)] Even Sn doping reveals presence of IP in $2\leq x\leq 5$ range.[(4)]

These research works provide new knowledge crucial for developing electronic storage technologies. These glasses are suitable for potential applications in PCM technologies and contribute toward refining electronic memory devices.[(1–4)]

References

1. Diptoshi Roy, B. Tanujit, K. B. Jagannatha, S. Asokan & Chandasree Das, *IEEE Transactions on Electron Devices (2021)* https://doi.org/10.1109/TED.2021.3051925
2. Jagannatha K B, Diptoshi Roy & Chandasree Das, *Journal of Non-Crystalline Solids (2020)* https://doi.org/10.1016/j.jnoncrysol.2020.120196
3. Diptoshi Roy, B. Tanujit, Jagannatha K. B, G. Sreevidya Varma, S. Asokan & Chandasree Das, *Journal of Non-Crystalline Solids (2020)* https://doi.org/10.1016/j.jnoncrysol.2019.119863
4. Jagannatha K B, B. Tanujit, Diptoshi Roy, S. Asokan & Chandasree Das, *Journal of Non-Crystalline Solids (2021)* https://doi.org/10.1016/j.jnoncrysol.2021.121311.

Phys. Chem. Glasses: Eur. J. Glass Sci. Technol. B, 2025, **66**(4), 158 DOI: 10.13036/17533562.66.4.AM46

Structural effects of controlled iron redox in obsidian-like glass systems

Andrew Parnapy & Doris Möncke

New York State College of Ceramics at Alfred University, Alfred, New York, 14802, USA

Corresponding author email: ajp16@alfred.edu

Obsidian is a naturally formed, silica rich glass that can contain relatively high amounts of iron. Obsidian and materials like obsidian (molten rock melts), have been used by humans for thousands of years as a building material. Specifically, in the pre-Viking hillforts of Scandinavia, vitreous rock melts were used as a mortar-like material in gaps of rock wall emplacements surrounding the hillforts. Modern applications can be seen in surgical blades worldwide or insulation where a more basaltic composition is used (i.e. Rockwool in Denmark).

Intervalence charge transfers (IV-CT) between the Iron (II) and Iron (III) are partially responsible for the coloration of black obsidian. By swapping Fe_2O_3 for $FeCO_3$, in a traditional melt quench, we hoped to control the ratio of Fe^{2+} and Fe^{3+} in obsidian-like melts. IV-CT transitions obscure the distinct d-d transition of Fe^{2+} and Fe^{3+}. However, in addition to different absorption spectra, divalent and trivalent iron species were expected to occupy different roles in the glass network. Fe^{2+} is expected to act more similar to CaO as a modifier, while $[Fe_3O_4]^-$ might play a role, more similar to tetrahedrally coordinated aluminate species, as a conditional network former. Controlling the ratio of these iron species meant we could also control the connectivity of the melts and thus the thermal properties exhibited.

Glasses of the $54SiO_2$-$13CaO$-$9Al_2O_3$-$10Fe_2O_3$-$5NaO$ (all other components <5%) composition were prepared with varying iron II/III ratios analyzed by differential scanning calorimetry and dilatometry to determine the thermal properties. Raman and UV/Vis spectroscopies (the latter with integrated sphere) were used to obtain information on structural changes and iron speciation. Systematic changes were observed in the thermal data as a function of the iron redox ratio.

References

1. Sjöblom, R.; Hjärthner-Holdar, E.; Pearce, C. I.; Weaver, J. L.; Ogenhall, E.; McCloy, J. S.; Marcial, J.; Vicenzi, E. P.; Schweiger, M. J.; Kruger, A. A. Assessment of the Reason for the Vitrification of a Wall at a Hillfort. The Example of Broborg in Sweden. *Journal of Archaeological Science: Reports* **2022**, *43*, 103459.
2. Losq, C. L.; Berry, A. J.; Kendrick, M. A.; Neuville, D. R.; O'Neill, H. S. C. Determination of the Oxidation State of Iron in Mid-Ocean Ridge Basalt Glasses by Raman Spectroscopy. *The American Mineralogist* **2019**, *104* (7), 1032–1042.

Phys. Chem. Glasses: Eur. J. Glass Sci. Technol. B, 2025, **66**(4), 159 DOI: 10.13036/17533562.66.4.AM47

Foam glass manufacturing as sustainable reuse of waste glass

Dr Arjen Steiner

SGGC GmbH, Bochumerstr. 217, 45886 Gelsenkirchen, Germany

Corresponding author email: a.steiner@sggc.com

Foam glass or cellular glass, is a lightweight, porous glass material with a rigid structure. It is made by heating a mixture of powdered glass and one or more additives/ foaming agents and is used in different applications. From soil exchange and road construction (foam glass gravel), or being a high value add to cementitious binding products (foam glass granulate) to being a high end, dimensionally stable, block or plate for heat insulation (foam glass blocks).

Many million tons of glass are recycled and reused in the world's glass industry. The smaller fractions of the recycled glass have limited use and are often disposed off in landfills. As the starting material for foam glass is mainly a glass powder these fractions can be a perfect source.

The presentation will show the path from waste glass recycling to a high end products showing the advantages manufacturing foam glass products. The requirements for the raw materials as well as foaming methods will be presented showing the advantages and disadvantaged of each method. The different production processes will be discussed and examples for the applications given.

Phys. Chem. Glasses: Eur. J. Glass Sci. Technol. B, 2025, **66**(4), 160 DOI: 10.13036/17533562.66.4.AM48

An increasing demand for materials recovered from existing buildings, a case study focused on glass materials and their second uses

Graeme DeBrincat[1] *& Florence Wu*[1]

[1]*Ove Arup & Partners, 8 Fitzroy St, London W1T 4BJ, UK*

Corresponding author email: graeme.debrincat@arup.com

Applying circular economy principles to building and construction materials is finally becoming a common approach in the design of new and existing buildings. Driven by new legislation and regulation, construction professionals are demanding materials that have a low carbon footprint, created with high recycled content and as low as possible energy consumption. As such, a healthy demand exists for the materials that can be collected from existing buildings to feed the supply into the manufacturing industries for use as alternative raw materials, changing away from the business-as-usual approach that utilises virgin raw materials.

Architectural glass is key material in the efforts to decarbonise the not only the glass making industries but wider material manufacture. Whilst the demand for these materials is high from the glass industry, the supply of suitable materials suffers significantly from low take up of careful dismantle and collection approaches to demolition and renovation, slowed further by a lack of technology and appropriate tools and techniques to fulfil the circular ambitions of both projects and the wider industry.

Flat glass manufacturers are rightly focused on optimising their operational carbon, currently that focus prioritises new furnace technology, incorporating fuel switching, waste heat recovery and carbon capture. Increasing the use of recycled content with high quality, sorted glass cullet, collected from existing buildings can play a crucial role in reducing the energy demand and carbon footprint of new glass products in the short term until these other technologies come online.

The embodied carbon of glass and the benefits of increased cullet use in the production of new flat glass is widely understood and demand for these materials far outstrips the supply of quality cullet materials. Research by manufacturers have shown that both higher percentages of overall cullet and in particular post-consumer cullet is feasible. Despite understanding the benefits, there are both real and perceived limitations to the implementation of post-consumer cullet in flat glass. Glass needs to be suitably collected and processed not only for the glass making industries but for other manufacturing applications as well.

Understanding the available glass types, their dismantling and separation requirements, requires new tools and technology to assess their qualities and compositions which will allow planning for the next use of these materials. Through identification of the constituent materials in architectural glass products and a through knowledge of uses of such glass materials, it may be possible to envisage a future where all glass materials are treated as essential and valuable raw materials for a wide range of new materials that have a significantly lower environmental impact then their current alternatives.

Phys. Chem. Glasses: Eur. J. Glass Sci. Technol. B, 2025, **66**(4), 161 DOI: 10.13036/17533562.66.4.AM49

Volta Project – decarbonisation of flat glass manufacturing

Niels Schreuder

AGC Glass Europe, Belgium

Corresponding author email: Niels.Schreuder@agc.com

In February 2025, the official inauguration of the Volta production line at the AGC Barevka plant in Teplice (Czech Republic), marked a milestone moment in the flat glass industry's strive for decarbonisation.

This breakthrough pilot project aims to significantly reduce CO_2 emissions and support the transition to sustainable flat glass production. The pilot furnace uses an innovative glass production technology that has never been used in the world before.

The Volta R&D project, a cooperation between AGC and Saint-Gobain, combines the technical know-how and innovative approaches of both companies. The two companies share the ambition to significantly reduce CO2 emissions and contribute to a faster transition towards clean technologies.

The project is supported by the EU Innovation Fund and includes the reconstruction of the production line of AGC Barevka (Czech Republic). The line will be 50% electrified and 50% fired by a combination of oxygen and gas, allowing for a significant reduction in direct carbon emissions. The new technology also focuses on circularity as the Volta hybrid melting furnace will operate at a higher ratio of recycled materials, much higher than the industry standards.

The pilot is done on this mid-sized furnace producing patterned glass and is an excellent test bed for both Saint-Gobain and AGC to test and validate the combined production technology knowledge in reality. The hybrid furnace is expected to reduce the pilot site's scope 1 and 3 emissions by 75% compared to the reference scenario.

Niels Schreuder is AGC Glass Europe's Public Affairs & Communication Director and will share AGC's views on circularity and decarbonisation.

Funded by the European Union
Emissions Trading System
Innovation Fund

References:

1. https://www.agc-glass.eu/en/sustainability/hybrid-mid-sized-pilot-furnace-for-flat-glass

Phys. Chem. Glasses: Eur. J. Glass Sci. Technol. B, 2025, **66**(4), 162 DOI: 10.13036/17533562.66.4.AM50

Making glass the clear choice for a sustainable future

Vasco Sousa

ISAG Instituto Superior de Administração e Gestão, Porto, Portugal

Corresponding author: vsousa@picvisa.com

PICVISA is a technology-driven company specializing in advanced technological solutions which integrate robotics, artificial intelligence and machine vision. We will explore our design philosophy and manufacturing capabilities of our innovative equipment for automation of existing or new glass plants using proprietary software to analyze material composition, shape, and colour, and the process of selecting and classification valuable materials for onward use or recycling. PICVISA's AI technologies for glass recovery will be detailed along with our function as an R&D focused company, seeking to drive innovation forward with our industrial and academic partners around the world.Sustainability in glass manufacturing is not as simple as it may seem due to challenges of diverse and complex material types, cleanliness and lack of a unified industry wide strategy or legislation.

Phys. Chem. Glasses: Eur. J. Glass Sci. Technol. B, 2025, **66**(4), 163 DOI: 10.13036/17533562.66.4.AM51

Challenges and early successes of running a 30 tonnes per day pilot line

Rob Ireson, Bridget Stewart & Daniel Backhouse

Glass Futures, James Roby Way, St Helens, Merseyside, WA9 5DT, UK

Corresponding author email: bridget.stewart@glass-futures.org

Glass Futures is a not-for-profit, membership-based, research organisation, based at its Global Centre of Excellence in St Helens, UK. The remit of the organisation is to enable its members from across the glass supply chain and beyond to collaborate in areas which affect all parties, with a particular focus on decarbonisation of the glass-making process and its upstream and downstream activities.

At the heart of the new Glass Futures facility is a 30 tonnes/day pilot-scale glass furnace, which started up in June 2025. The facility has been designed to enable the industry to develop and trial new technologies at an industrially relevant scale, without risk to their commercial manufacturing assets, thus providing increased confidence for manufacturers looking to invest in low-carbon technologies such as alternative fuels, CCUS technologies and new raw materials.

The operation of such a globally unique research asset is a journey into the unknown, with procedures and techniques which are common in the glass manufacturing sector having to be adapted to manage experimental programmes rather than manufacture for sale. In this paper, an overview of the facility is provided, the initial challenges and successes of operating such an asset are discussed and key learning points are highlighted.

Phys. Chem. Glasses: Eur. J. Glass Sci. Technol. B, 2025, **66**(4), 164 DOI: 10.13036/17533562.66.4.AM52

The next generation of industrial glass furnaces

Owen McGann & Martyn Marshall

Glass Technology Services Ltd, 9 Churchill Way, Chapeltown, Sheffield, S35 2PY, UK

Corresponding author email: o.mcgann@glass-ts.com

For over 250 years, glass has been produced on an industrial scale, beginning with the manufacture of glass bottles and crown glass windows in coal-fired furnaces in Britain. Over that time, the technology used for large-scale glass melting has evolved—most notably through the invention of regenerative furnaces and the development of high-silica and high-alumina refractory materials, which have defined the shape and operation of industrial glass furnaces for most of the past 150 years.

Today, the glass industry faces increasing pressure from rising energy costs and the imperative to reduce CO_2 emissions. This has accelerated efforts to innovate, particularly through the adoption of alternative heat sources and the optimisation of furnace operations. While such improvements offer significant benefits, the question remains: could more radical changes in furnace design and operation deliver even greater gains?

Drawing on the cross-industry expertise and long history of Glass Technology Services, this talk will review potential 'next generation' technologies. These include approaches explored in past decades that were not adopted at the time (e.g. segmented melting), technologies developed in adjacent fields that may now hold relevance for mainstream glass production (e.g. joule-effect and plasma melting), and emerging innovations that have yet to be applied. The talk will assess these technologies in the context of the modern challenges facing the glass industry and offer insights into what may define the next generation of industrial glass furnaces.

Phys. Chem. Glasses: Eur. J. Glass Sci. Technol. B, 2025, **66**(4), 165 DOI: 10.13036/17533562.66.4.AM53

Beneficiation of ash and slag for sustainable glass production

Sabrin A. Samad, Anthony Bell, Wei Deng, Feroz Kabir Kazi, Hywel Jones & Paul A. Bingham

Materials and Engineering Research Institute, Sheffield Hallam University, Sheffield, S1 1WB, UK

Corresponding author email: s.samad@shu.ac.uk

The glass industry, one of the largest globally with production reaching 39·12 million tons in 2021 and a market value exceeding €28 billion, continues to face challenges related to high energy consumption, CO_2 emissions and competition from alternative materials. In this context, biomass ash and slag present a promising opportunity as an alternative raw material capable of driving significant progress towards decarbonising the glass industry. This study investigates processes and approaches to beneficiate ash and slag from gasification process to improve their suitability for industrial glass production. The biomass ashes and slag were characterised by x-ray fluorescence (XRF), x-ray diffraction (XRD), and Ultimate analysis (C,N,H,S) to determine ash composition. The results revealed that the ash samples contain 20–30% of carbon of the total ashes along with considerable amounts of chlorine, sulphur and heavy elements while slag samples were low in carbon, chlorine, sulphur and heavy elements. Both samples were then subjected to water leaching at 90°C, 75°C, 50°C and room temperatures for 1, 2 and 3 h, followed by analysis of leachates using ICP-MS, pH and conductivity measurements. XRF and ICP results confirm that the leaching process significantly reduced chlorine and sulphur levels, indicating high solubility and effective removal. Subsequent combustion of the samples reduced the carbon content from 20-30% to below 1% as confirmed via TGA and C,N,H,S analysis. Magnetic separation was employed to reduce iron levels and then the samples were sieved into respective size fractions, followed by incorporation into green soda–lime–silica (SLS) glass at a 5% substitution level using both raw and beneficiated ash and slag. The resulting glasses were then analysed and compared to benchmark green glass using XRF and UV-VIS analysis. The findings demonstrate that beneficiation route effectively removes impurities and inhomogeneity, offering a more resource-efficient and economically viable pathway aligned with circular economy for glass manufacturing.

Phys. Chem. Glasses: Eur. J. Glass Sci. Technol. B, 2025, **66**(4), 166 DOI: 10.13036/17533562.66.4.AM54

Reducing energy consumption and carbon dioxide emissions from glass manufacture

Ebele A. Ahizi, Wei Deng, Feroz Kabir & Paul A. Bingham

Materials and Engineering Research Institute, Sheffield Hallam University, City Campus, Howard Street, Sheffield, S1 1WB, UK

Corresponding author email: e.ahizi@shu.ac.uk

There is a global effort to reduce the energy demand and carbon dioxide emissions from glass manufacture to meet the current net zero targets for a sustainable environment. Alternative raw materials are being sourced which have the capacity to reduce energy usage and limit CO_2 emissions from glass manufacture. Among these is wollastonite ($CaSiO_3$), which has been identified as a rich source of calcium oxide potentially capable of partially or completely replacing limestone in glass batches whilst also delivering faster melting kinetics, leading to energy saving during glass manufacturing. Different lab-scale experimental analyses have been carried out to ascertain the viability of wollastonite as a limestone substitute in soda-lime-silica (SLS) glass manufacture. Among the analyses carried out were glass melting including stages-of-melting study, with XRF, XRD, DSC/TGA, high temperature glass monitoring, and micro-CT. Following the findings so far, when a batch containing wollastonite as the only calcium oxide source in SLS glass is compared to a benchmark glass composition containing limestone ($CaCO_3$) as the calcium oxide source, wollastonite maintained faster reaction kinetics, lesser number of bubbles at a comparable temperature. These and more research discoveries point to its ability to have lower energy demand during the batch reaction processes and consequently, lower melting energies. Further analyses are underway to quantify melting behaviour and potential energy savings.

Phys. Chem. Glasses: Eur. J. Glass Sci. Technol. B, 2025, **66**(4), 167 DOI: 10.13036/17533562.66.4.AM55

Vitreous radioactive wasteform development: an overview

Rachel Crawford,[1,2] *Claire Corkhill,*[2] *Russell J. Hand,*[2] *Clare Thorpe*[2] *& Paul A. Bingham*[1]

[1] *Sheffield Hallam University, City Campus, Howard Street, Sheffield, S1 1WB, UK*
[2] *Immobilisation Science Laboratory, University of Sheffield, Western Bank, Sheffield, S10 2TN, UK*

Corresponding author email: rachel.crawford@shu.ac.uk

Nuclear energy is used globally and is promoted in efforts to decarbonise the generation of electrical energy and is well established in the energy sector. Across the lifespan of a reactor, several waste streams are produced and must be managed with care and consideration across vast timescales. As such, the long-term management and disposal of radioactive wastes is discussed in national policies and has led to many international collaborations, across industry and academia.

Immobilisation via vitrification is a method for radioactive waste management utilised in many nations. For example, both the USA and the UK vitrify both high and intermediate level wastes, with further work ongoing to develop additional technologies to vitrify complex and varied waste streams.[1,2] *Vitrification has many benefits, volume reducing waste whilst also producing a durable, radiation tolerant material, stable across geological timescales.*

Wasteform development requires thorough investigations of waste envelopes, analysis and characterisation of developed waste compositions. These investigations are instrumental in forming a safety case, determining if a proposed wasteform composition is safe for long term geological disposal.[3] *Safety case development is a crucial part of implementing waste disposal policies. Understanding the long-term behaviour of waste forms and ensuring that geological disposal reduces and inhibits the release of radionuclides into the environment.*

This presentation outlines varying stages of wasteform development, from discussions surrounding waste envelopes, wasteform durability assessments and interactions with barrier materials. These areas are crucial to the overall understanding of a vitrified wasteform, with each stage providing complex research challenges.

References

1. Corkhill, Claire & Hyatt, Neil. (2018). Nuclear Waste Management. 10.1088/978-0-7503-1638-5.
2. José Marcial, Brian J. Riley, Albert A. Kruger, Charmayne E. Lonergan, John D. Vienna, Hanford low-activity waste vitrification: A review, Journal of Hazardous Materials, Volume 461, 2024, 132437, ISSN 0304-3894, https://doi.org/10.1016/j.jhazmat.2023.132437.
3. Ojovan, Michael. (2023). Nuclear Waste Disposal. Encyclopedia. 3. 419-429. 10.3390/encyclopedia3020028.

Phys. Chem. Glasses: Eur. J. Glass Sci. Technol. B, 2025, **66**(4), 168 DOI: 10.13036/17533562.66.4.AM56

Sodalime and volcanic ash alkali activated foams for organic dye sorption

G. Tameni,[1] *F. Lanero,*[1] *E. Russo,*[1] *L. Moro, P. Sgarbossa,*[1] *L. Contrafatto*[2] *& E. Bernardo*[1]

[1] *Department of Industrial Engineering, University of Padova, Italy*
[2] *Department of Civil Engineering and Architecture, University of Catania, Italy*

Corresponding author email: giulia.tameni@phd.unipd.it

Soda-lime waste glass represents a material that holds potential for valorization in the alkali activation process. Similarly, volcanic ash, which constitutes a waste material, can be utilized in the production of cementitious materials or alkali activated materials. Fine glass powder (GP, ϕ<32 μm) and volcanic ash powder (VA, ϕ<75 μm), with a different ash-glass ratio (VA:GP=0:100, 25:75, 50:50, 75:25, 100:0), are suspended in a mild alkaline solution (NaOH, 3 mol/L). Liquid-to-solid ratio varies between 0·35, 0·39, and 0·42. The obtained slurries are consolidated in an oven for 3 days at 75°C prior to testing. Both solid and foamed materials have been produced.[(1)] *The solid materials exhibit characteristics comparable to those of construction materials in terms of the compressive strength-to-density ratio. Foamed materials have also been produced using 'direct foaming' process techniques, incorporating cost-effective and available foaming agents such as sodium perborate monohydrate (SP, 1 wt%) and sodium dodecyl sulphate (SDS, 1 wt%).*[(2)] *The foams exhibit an excellent thermal conductivity-to-density ratio and can be used as highly effective thermal insulators. The foams are also functionalised in order to obtain magnetic object through the addition of Fe_3O_4 covered by a SiO_2 layer in concentration of 1, 5 and 10 wt%. Moreover, natural zeolites present in alkali activated materials have been utilised to conduct adsorption test on two different organic pollutant dyes: methylene blue ($C_{16}H_8ClN_3S$) at 10 mg/L and rhodamine-B ($C_{28}H_{31}ClN_2O_3$) at 4 mg/L. The preliminary results show that alkali activated functionalised foams have promising photo-adsorption capabilities.*

References

1. Bernardo, E. & Elsayed, H. & Mazzi, A. & Tameni, G. & Gazzo, S. & Contrafatto, L. Double-life susta*inable construction materials from alkali activation of volcanic ash/discarded glass mixture 2022, Construction and Building Materials 359, 129540,* https://doi.org/10.1016/j.conbuildmat.2022.129540
2. Tameni, G. & Lago, D. & Kaňková, H. & Buňová, L. & Kraxner, J. & Galusek, D. & Daeson, D.M. & Ashbrook, S.E. & Bernardo, E., Alkaline attack of boro-a*lumino-silicate glass: New insights of the molecular mechanism of cold consolidation and new applications 2025, Open Ceramics 21 100726,* https://doi.org/10.1016/j.oceram.2024.100726

Phys. Chem. Glasses: Eur. J. Glass Sci. Technol. B, 2025, **66**(4), 169 DOI: 10.13036/17533562.66.4.AM57

Academia and industry collaboration for a sustainable glass future

Rob Werner & Chris Holcroft

Glass Technology Services, 9 Churchill Way, Chapeltown, Sheffield S35 2PY, UK

Corresponding author email: r.werner@glass-ts.com

GTS has a long history of bridging the gap between academic theory and industrial implementation. In this presentation we will share some examples of this collaboration and highlight the benefits this can bring in achieving sustainability goals in the glass industry. Over recent years we have worked with many UK and global academic institutions on projects to increase recycled content of glass, cement and other materials, develop glasses for dental and orthopedic use, and vitrify hazardous waste material. We are also proud to support and host PhD, undergraduate and degree apprentice students, helping to create a sustainable skilled workforce in the glass and materials science sector in the future.

Phys. Chem. Glasses: Eur. J. Glass Sci. Technol. B, 2025, **66**(4), 170 DOI: 10.13036/17533562.66.4.AM58

Summarising the SGT Sustainability Sessions and formulating a plan for a low carbon future

Katrina Skerratt-Love[1]

[1] *Department of Materials, Design and Manufacturing, University of Liverpool, Liverpool, L69 3GH, UK*

Corresponding authors email: Katrina.skerratt-love@liverpool.ac.uk

Sustainability can mean many things according to a perspective or motivation, yet as an industry we have relied on innovation in product design, composition, and production methodology to reduce our carbon footprint, but is reducing our carbon footprint the same as sustainability or even circularity. The SGT Sustainability sessions have been structured to offer attendees the opportunity of understanding how the industry has evolved, the legislation which governs, and some potential solutions for achieving our challenging carbon reduction targets. Our session will summarise the state of the global glass industry using the presentation offered by industry experts and attendee interactions to create the foundations of a market assessment, offering both attendees and the wider audience the opportunity to consider how the industry could evolve with the correct legislation, technical solutions, and financial motivation. We will not just focus on improving cullet quantity and quality but consider other uses which offer low carbon markets for material often landfilled. Improving collaboration between Academia and Industry is imperative, encouraging greater interaction can be challenging when we are unclear of the desired outcome, understanding the complexity and challenges of each glass sector will help frame a strategy of greater cooperation where improving the performance of our industry and preparing for a more sustainable future is prioritised.

Phys. Chem. Glasses: Eur. J. Glass Sci. Technol. B, 2025, **66**(4), 171 DOI: 10.13036/17533562.66.4.AM59

Mid-infrared photonics: a new frontier in disease screening and diagnosis using chalcogenide glass fibreoptics

Angela B. Seddon,[1] *Jessica Grant,*[1] *Roberta Schiemer,*[2] *Waziri Addo-Yobo, David Furniss,*[1] *Radu Boitor,*[3] *Ioan Notingher,*[3] *Sendy Phang,*[1] *Mark C. Farries,*[1] *Nia W. Jones,*[4] *William Atiomo*[5] *& Ketankumar B. Gajjar*[2]

[1] *Mid-Infrared Photonics Group, George Green Institute for Electromagnetics' Research, Faculty of Engineering, University of Nottingham, Nottingham NG7 2RD, UK*
[2] *Division of Child Health, Obstetrics and Gynaecology, University of Nottingham, Nottingham NG5 1PB, UK*
[3] *School of Physics and Astronomy, Faculty of Science, University of Nottingham, Nottingham NG7 2RD UK*
[4] *Division of Lifespan and Population Health, University of Nottingham NG5 1PB, UK*
[5] *College of Medicine, Mohammed Bin Rasid University of Medicine and Health Sciences (MBRU), Dubai P.O. Box 505055, United Arab Emirates*

Corresponding author email: angela.seddon@nottingham.ac.uk

Introduction: Mid-infrared (MIR) spectroscopy has potential as a new pathology for cancer screening and diagnosis. The MIR spectral region (wavelengths 3–50 mm [BS ISO 20473: 2007]) corresponds to the fundamental vibrational frequencies of molecular bonds. MIR light probes these to provide a molecular "fingerprint" of biological samples. MIR light detects subtle biochemical changes in tissues/biofluids indicative of disease, like changes in protein, lipid and nucleic acid content. Cancerous tissue/bio-fluids, versus healthy tissue/ biofluids have different molecular compositions that MIR spectroscopy can differentiate, helping to identify disease at an early stage. We are working on endometrial cancer (EC) detection. EC detection at Stage-4 has ca. 1/6 the survival rate of EC detection at Stage-1, and ~3x the treatment costs, hence the need for early EC detection.[1]

Screening EC: currently there is no UK NHS (National Health Service) screening for EC: a simple blood test is attractive. Working retrospectively with thawed frozen blood samples, it is found that a combined ATR-FTIR (attenuated total reflectance, Fourier transform infrared)-Raman spectroscopy approach exhibits the highest classification accuracy of 86% for 'wet' blood-plasma samples. (Note, Quasar software with supervised SVM (support vector machine) statistics is applied; all classifiers are trained on a reduced dataset of 70% training-spectra with 30% test-spectra and 10-fold cross-validation to prevent overfitting.)

In vivo diagnosis of EC: we are developing spectroscopic sensing of cancer based on new chalcogenide glass, MIR fibreoptics for a new MIR endoscopy, enabling new MIR 'optical biopsies'. Practical deployment requires the means to generate bright MIR light across a broad MIR spectrum, and to route it to where needed. We have developed: (i) low optical loss, non-cytotoxic fibre for routeing;[2] *(ii) the first nonlinear, MIR super-continuum (SC) 'rainbow' fibre laser (with DTU, Denmark) to deliver bright light across the MIR spectrum;*[3] *and (iii) the first continuous wave, room temperature fibre laser operating at >4 mm wavelength, with potential to pump the MIR SC rainbow source.*[4] *An update will be presented.*

Reference

1. Cancer Research UK, Womb cancer survival: Pennington, M. *et al. PLOS One, 2016*, Endometrial Cancer (thelancet.com), https://doi.org/10.1371/journal.pone.0165539
2. Evaluating the cytotoxicity of Ge–Sb–Se chalcogenide glass optical fibres on 3T3 mouse fibroblasts. Mabwa D, Kubiena T, Parnell H, et al. RSC Adv. 2021;11(15), 8682–8693.
3. Mid-infrared supercontinuum covering the 1.4–13.3 μm molecular fingerprint region using ultra-high NA chalcogenide step-index fibre. Petersen, C., Møller, U., Kubat, I. et al. Nature Photon 8, 830–834 (2014). https://doi.org/10.1038/nphoton.2014.213
4. Room temperature mid-infrared fiber lasing beyond 5 μm in chalcogenide glass small-core step index fiber. J. J. Nunes, Ł. Sojka, R. W. Crane, D. Furniss, Z. Q. Tang, D. Mabwa, B. Xiao, T. M. Benson, M. Farries, N. Kalfagiannis, E. Barney, S. Phang, A. B. Seddon, and S. Sujecki, Opt. Lett. 46, 3504-3507 (2021).

Phys. Chem. Glasses: Eur. J. Glass Sci. Technol. B, 2025, **66**(4), 172 DOI: 10.13036/17533562.66.4.AM60

Optimisation of Er^{3+}/Yb^{3+} co-doped phosphate-based laser glasses for enhanced laser performance

Ahmet Caner Kayaalp,[1] *Ali Erçin Ersundu,*[2] *Alphan Sennaroğlu*[3] *& Miray Çelikbilek Ersundu*[2]

[1] *Şişecam Science, Technology and Design Center, Çayırova, Şişecam Yolu St., 41420, Kocaeli, Turkiye*
[2] *Yildiz Technical University, Faculty of Chemical and Metallurgical Engineering, Department of Metallurgical and Materials Engineering, Glass Research and Development Laboratory, Istanbul, 34220, Turkiye*
[3] *Koç University Surface Science and Technology Center (KUYTAM), Rumelifeneri, Istanbul 34450, Türkiye*

Corresponding author email: cakayaalp@sisecam.com

Phosphate glasses have found wide use in the field of photonics over the past two decades due to their high solubility rate (up to 10^{21} ions/cm^3) of rare-earth elements, low phonon energy and wide transmittance range, which are essential for laser gain materials. Er^{3+} and Yb^{3+} co-doped phosphate laser glasses, emitting eye-safe radiation at 1·54 μm, come into prominence and are used in range finders, target locking systems and medical lasers for dermatology. This study focuses on the development of a highly Yb^{3+} doped glass composition, up to 12 mol%, with low crystallisation tendency and excellent thermomechanical properties expected from laser glasses, as well as the investigation of the effect of varying Er^{3+} and Yb^{3+} doping ratios on laser properties. During the glass synthesis, OH^- ions were reduced to 7 ppm using a two-step melting process under an inert atmosphere. The OH^- absorption coefficient was optimised with varying heat-treatment temperatures and times, improving the emission properties.

Chemical composition of the glasses was determined by XRF analysis, confirming its compliance with the nominal composition. Transmittance and absorption properties over a wide range were investigated by UV-Vis-NIR and FTIR analysis. Structure of the samples were investigated FTIR analysis. Glass transition temperature, coefficient of thermal expansion, and viscosity values were measured and evaluated based on the increasing Yb^{3+} ratio. Refractive index analyses were found to show an increase with increasing Yb_2O_3 content. Photoluminescence measurements under 976 nm excitation were performed to determine the emission intensity and lifetime values of the glasses both before and after dehydration. The laser threshold, output power and slope efficiency were measured to evaluate the laser performance efficiency of the samples. Based on the results, the ideal Er^{3+} and Yb^{3+} doping ratios for achieving the highest laser performance were determined, energy transfer between Er^{3+} and Yb^{3+} was defined, and effect of dehydration on laser performance was evaluated.

The authors acknowledge the Scientific & Technological Research Council of Turkey (Tübitak) for the financial support under the project numbered 119C205.

Phys. Chem. Glasses: Eur. J. Glass Sci. Technol. B, 2025, **66**(4), 173 DOI: 10.13036/17533562.66.4.AM61

Structure and redox state of the sulphur-containing glass melts

Justyna Sułowska

Faculty of Materials Science and Ceramics, AGH University of Krakow, Al. Mickiewicza 30, 30-059 Krakow, Poland

Corresponding author email: sulowska@agh.edu.pl

Sulphur is a poorly glass-compatible element. It can be incorporated into silicate and borosilicate glass structures at up to 1 wt% of SO_3 equivalent. More than 1 wt% SO_3 causes the formation of a separate water-soluble (yellow) phase that contains alkali sulphates.[1] *According to the literature,*[2] *the solubility of sulphur in the structure of phosphate glasses is higher than that in silicate or borosilicate glasses. The volatilisation of SO_2 gas due to the synthesis of sulphate glasses is the largest problem.*

The aim of the study was to analyse the redox state and structure of glasses from the SiO_2–P_2O_5–K_2O–MgO–SO_3–ZnO system. The glasses were obtained under reducing conditions. The bulk composition of the glasses was confirmed by x-ray fluorescence spectroscopy. The amorphous nature of the samples was verified using XRD analysis. The structure of the materials was studied with the spectroscopic techniques: ^{29}Si and ^{31}P MAS NMR, and Raman spectroscopy.

Acknowledgments

Research project supported by program "Excellence initiative – research university" for the AGH University, grant ID 8874.

References

1. M.I. Ojovan, W.E. Lee, "An introduction to nuclear waste immobilization". Elsevier. Amsterdam, The Netherlands, 2005.
2. K. Ghosh, G.K. DasMohapatra, N. Soodbiswas, "Glass formation in K2SO4-CaO-P2O5 system". Phys. Chem. Glas. 44 (2003) 313–318.

Phys. Chem. Glasses: Eur. J. Glass Sci. Technol. B, 2025, **66**(4), 174 DOI: 10.13036/17533562.66.4.AM62

Hollow core fibres: when nothing is better than glass!

Francesco Poletti[1,2]

[1] *Optoelectronics Research Centre, University of Southampton, UK*
[2] *Microsoft Azure Fiber, Romsey, UK*

Corresponding author email: fp@soton.ac.uk

Long before the development of ultra-low-loss optical glasses in the 1970s, hollow waveguides – essentially pipes of air – were the only viable means of guiding electromagnetic waves and transmitting information.[(1–3)] *The invention of ultrapure fused silica fibres*[(4)] *marked a turning point, enabling the birth of modern optical communications. Despite extensive efforts to develop alternative glass compositions, only fused silica has met the stringent attenuation requirements for long-distance data transmission. For the past 50 years, it has remained the material of choice, enabling the fabrication of over 5 billion kilometres of optical fibre that now span the globe.*

Now, a new generation of hollow core optical fibres (HCFs) is challenging this long-standing paradigm.[(5)] *These fibres have recently achieved attenuation levels lower than what is fundamentally possible with any solid glass.*[(6)] *With light confined primarily to the air core and interacting with the glass cladding at levels as low as 50 parts per million, the need for ultra-pure glass is dramatically reduced.*

This breakthrough invites a re-examination of glass families previously deemed unsuitable for optical communications. Materials such as soda–lime–silica glass, borosilicates, alumino-silicates and lead silicates – once dismissed due to their higher intrinsic losses – may now find new relevance. These glasses offer potential cost reductions and greater design flexibility. However, it is paramount that any alternative glass also provides the mechanical strength and durability required for fibres to survive decades of deployment in the field.

Could this be the beginning of a renaissance in optical glass research? This talk will explore the historical evolution of waveguiding, the physics behind modern HCFs, and the exciting opportunities that arise when "nothing" – air – becomes the optimal medium for light.

References

1. Rayleigh, Lord. On the passage of electric waves through tubes, or the vibrations of dielectric cylinders . *The London, Edinburgh, and Dublin Philosophical Magazine and Journal of Science* **43**, 125–132 (1897).
2. Warters, W. D. WT4 Millimeter Waveguide System: Introduction. *Bell System Technical Journal* **56**, 1873–1897 (1977).
3. Barrow, W. L. Transmission of electromagnetic waves in hollow tubes of metal. *Proceedings of the IEEE* **72**, (1984).
4. Keck, D. B., Maurer, R. D. & Schultz, P. C. On the ultimate lower limit of attenuation in glass optical waveguides. *Appl Phys Lett* **22**, (1973).
5. Poletti, F. Nested antiresonant nodeless hollow core fiber. *Opt Express* **22**, (2014).
6. Chen, Y. *et al.* Hollow Core DNANF Optical Fiber with 0.11 dB/km Loss. *Optical Fiber Communication Conference (OFC) 2024, paper Th4A.8*

Phys. Chem. Glasses: Eur. J. Glass Sci. Technol. B, 2025, **66**(4), 175 DOI: 10.13036/17533562.66.4.AM63

Tailoring rapid dissolution of metaphosphate glasses for biocompatible applications

Lucas Greiner[1,2] *& Doris Möncke*[1]

[1] *New York State College of Ceramics at Alfred University, Alfred NY 14802, USA*
[2] *Missouri University of Science and Technology, Rolla MO 65409, USA*

Corresponding author email: llgb4b@mst.edu lg18@alfred.edu

Phosphate glasses continue to attract significant interest due to their low processing temperatures, biocompatibility and highly tunable dissolution behaviour, making them promising candidates for biomedical, agricultural and nuclear waste immobilisation applications. In this work, phosphate-based glasses were synthesised and tailored for controlled degradation, with a particular focus on rapid-dissolving compositions suitable for temporary biomedical scaffolds. Glasses within the ultra-, meta- and polyphosphate families, ranging from 45 to 55 mol% P_2O_5*, were fabricated and systematically studied.*

Static dissolution tests in deionised water were conducted to measure weight loss over time and inductively coupled plasma optical emission spectroscopy (ICP-OES) analysis was employed to track silicon concentrations from crucible dissolution. Structural and thermal properties governing the dissolution behaviour were investigated using Raman spectroscopy and differential scanning calorimetry coupled with thermogravimetric analysis (DSC-TGA). Results demonstrated that increasing CaO content led to higher glass transition temperatures (T_g) *and significantly improved thermal stability, with weight loss from TGA remaining below 1% up to 600°C for both metaphosphate and polyphosphate glasses.*

The incorporation of modifier oxides (e.g. Na_2O*,* K_2O*, CaO) are found to strongly influence phosphate chain stability, where increasing field strength modifiers, especially divalent cations like* Ca^{2+}*, promoted greater crosslinking leading to reduced solubility. Raman spectroscopy revealed notable structural changes and the transformation of phosphate chain structures as a function of both modifier content and phosphate concentration. These results highlight the critical role of composition in controlling glass network connectivity and hydrolytic stability.*

Overall, this study underscores the potential for compositional design to tailor the dissolution rates of phosphate glasses, enabling their targeted use in biocompatible implants, agricultural nutrient delivery, and durable wasteforms.

References

1. Bunker, B. C.; Arnold, G. W.; Wilder, J. A. Phosphate Glass Dissolution in Aqueous Solutions. *J. Non-Cryst. Solids* **1984**, *64* (3), 291–316. https://doi.org/10.1016/0022-3093(84)90184-4.
2. Brow, R. K.. *Review: The structure of simple phosphate glasses, Journal of Non-Crystalline Solids, Volumes 263-264, 2000, pgs 1-28, ISSN 0022-3093, http://doi.org/10.1016/S0022-3093(99)00620-1.*
3. Greiner, L. *Water Interactions with Silicate and Phosphate Glasses: Taking Advantage of Corrosion;* M.S. Thesis, Alfred University, 2024.

Phys. Chem. Glasses: Eur. J. Glass Sci. Technol. B, 2025, **66**(4), 176 DOI: 10.13036/17533562.66.4.AM64

Advances in glass-based space radiation shielding

Katrina Skerratt-Love,[1] *Matthew Gill*[2] *& Maulik Patel*[1]

[1] *Department of Materials, Design and Manufacturing, University of Liverpool, Liverpool, L69 3GH, UK*
[2] *Space Radiation Services, C/O Praxis, 1 Poultry, London, United Kingdom, EC2R 8EJ, UK*

Corresponding author email: Katrina.skerratt-love@liverpool.ac.uk

Space is a hostile environment for astronauts, where prolonged missions can result in severe short-and long-term health issues. One such factor is exposure to radiation characterised by galactic cosmic rays (GCR), alpha- and beta- particles, gamma rays, and neutrons.[(1)] *Efficient shielding against such radiation is very difficult, where different particles modify properties of materials differently but synergistically, hence shielding strategies need to be optimised for the environment.*

The ionizing radiation (IR) in space can cause ionization and displacement of atoms depending on the energy dosage and exposure duration leading to material property degradation.[(2)] *The GCR spectrum is made up of particles with a wide range of energies and mass, making it difficult to emulate the whole gamut of radiation environment in a terrestrial laboratory.*[(3)] *In addition, accumulation of damage in materials take a very long time in space, requiring accelerated methods to evaluate next generation materials for space application. The current approach is to use high energy accelerators, gamma and x-ray chambers to deliver radiation with specific energy and mass to evaluate material performance.*

Among various materials used in space technology, our specific interest is in glass and glass-ceramics for shielding applications to reduce the dose effects of IR from GCR.[(4)] *Glassy materials have been used as shuttle tiles, doors, windows, and as coatings.*[(5)] *The benefit of glass is its compositional flexibility, where it can be doped with specific elements to provide optimised properties, e.g. radiation, optical transparency. Recent studies have demonstrated that certain glass compositions based on oxides (B_2O_3, SiO_2, P_2O_5), and metal oxides (Sm_2O_3, GeO_2 and TeO_2) exhibit radiation resistance, when doped with either heavy metal oxides, lanthanides, or transitions metals. Two key challenges are designing glasses that are, optimized glass compositions for multiple, often competing properties and characterisation/quantification of radiation induced damage in an amorphous system.*

The current talk will provide an overview of radiation effects in glasses for space application and outline computational and experimental methods to overcome these challenges.

References

1. Scott, B. R. Radiation Toxicology, Ionizing and Nonionizing. in *Encyclopedia of Toxicology* 29–43 (Elsevier, 2014). doi:10.1016/B978-0-12-386454-3.00057-9.
2. Al-Buriahi, M. S., Kurtulus, R., Eke, C., Alomairy, S. & Olarinoye, I. O. An insight into advanced glass systems for radiation shielding applications: A review on different modifiers and heavy metal oxides-based glasses. *Heliyon* **10**, e40249 (2024).
3. Barthel, J. & Sarigul-Klijn, N. A review of radiation shielding needs and concepts for space voyages beyond Earth's magnetic influence. *Prog. Aerosp. Sci.* **110**, 100553 (2019).
4. Gohel, A. & Makwana, R. Multi-layered shielding materials for high energy space radiation. *Radiat. Phys. Chem.* **197**, 110131 (2022).
5. Tursunkhanova, R. B., Sergeev, V. P., Kalashnikov, M. P., Sergeev, O. V. & Neufeld, V. V. The effect of indium–tin oxide coatings on the formation of craters on glass surfaces under the impact of high-velocity microparticles. *Acta Astronaut.* **204**, 863–868 (2023).

Phys. Chem. Glasses: Eur. J. Glass Sci. Technol. B, 2025, **66**(4), 177 DOI: 10.13036/17533562.66.4.AM65

pH switchable bacteriocidal glasses

Robert Hill

DPSUDental Institute Queen Mary University of London, Mile End Road London E14NS, UK

Corresponding author email: r.hill@qmul.ac.uk

There is considerable interest in bacteriocidal glasses for paints coatings and implants. Issues like hospital acquired infections are a significant problem. One of the challenges of existing degradable glasses is that the release bacteriocidal ions is short lived. One way to address this problem is to release the bacteriocidal ions only in response to a bacterial infection. Most infections result in an acidic pH so we have designed glasses which are stable at neutral pH, but dissolve rapidly at acidic pHs.

This is achieved by incorporating acid hydrolysable bonds into a silicate glass network. These bonds can be Al–O–Si bonds or more preferably Zn–O–Si bonds. A new network connectivity model has been developed based on an understanding of glass structure that can be used to design suitable glass compositions. The model calculates network connectivity of the glass in the conventional way[1] *but then also calculates it again for acidic conditions assuming all the acid hydrolysable bonds are broken. In order to have a pH switchable glass we want a NC>2·6 under neutral conditions but an NC<2·0 under acidic conditions.*

Results will presented on: SiO_2*–ZnO–CaO–SrO–MgO–*CaF_2 *glasses, both as glass powders and as composites. Glass dissolution is followed by immersion in both 0·1M acetic acid pH4 and tris buffer pH 7 and monitor the ions in solution by inductively coupled optical emission spectroscopy.*

The application of the glasses in dentistry, medicine and paints will be briefly discussed.

References

1. Hill, R. An Alternative Explanation of the Bioactivity of Bioglasses. *J. Mater. Sci. Lett.*, 1996, **15**, 1122–25.

Phys. Chem. Glasses: Eur. J. Glass Sci. Technol. B, 2025, **66**(4), 178 DOI: 10.13036/17533562.66.4.AM66

At the intersection of art and science: exploring dichroic glass through reconstructions to use as an artistic medium

Zoe Bedford[1] *& Andreia Ruivo*[2]

[1] *Independent researcher, glass artist*
[2] *VICARTE – Vidro e Cerâmica para as Artes, Hangar III, NOVA School of Science and Technology (FCT NOVA), Portugal*

Corresponding email address: Zoe.bedford@artandconservation.com

Dichroic glass displays different colours when viewed in reflected versus transmitted light, representing a fascinating convergence of artistic expression and technical expertise. This optical phenomenon results from metallic nanoparticles dispersed throughout the glass matrix, creating works that transform with changing light conditions and viewer perspectives.

The study examines the compositions of three artifacts dated between approximately the fourth to the eighth century: the Lycurgus Cup, the Termancia Cup, and a fragment from near Newcastle, UK. The compositions were obtained from Robert H. Brills Chemical Analysis of Early Glasses Volumes I & II.[1] *By analysing these ancient examples, this research investigates chemical compositions, dichroism formation processes, and temperature effects on aesthetic outcomes. Advanced scientific techniques including UV-visible spectroscopy and X-ray fluorescence establish a comprehensive understanding of the underlying material properties and processes.*

Access to the glass composition marked the initial phase of our development process. As both artist and researcher, and scientist, we sought to explore the possibilities within the laboratory environment while understanding the unique properties of each glass—knowledge that would ultimately inform artistic application. To create authentic reconstructions, we designed a methodical laboratory approach for all three glass samples, carefully accounting for the constraints of modern equipment. Differential Scanning Calorimetry (DSC) guided our development of optimal kiln cycles. In subsequent phases, we analysed how varying melting temperatures affected the glasses and introduced controlled compositional modifications to better understand each element's contribution to the final product. Further analysis gave additional insight to how the dichroic effect was formed as the temperature came into effect. The outcome was successfully produced dichroic glasses.

References

1. Brill, R. H., *Chemical Analysis of Early Glasses Volumes I & II*. The Corning Museum of Glass, New York, 1999.

Phys. Chem. Glasses: Eur. J. Glass Sci. Technol. B, 2025, **66**(4), 179 DOI: 10.13036/17533562.66.4.AM67

High aspect ratio flat optical fibre for application in process control evaluation and health monitoring of high value composite assets

Christopher Holmes,[1] Pawel Maniewski,[1,2] Timothy Lee,[1] Robin Hartley,[3] Matthew Whitaker,[1] Bruno Moog,[1] Martynas Beresna[1] & Janice M. Dulieu-Barton[3]

[1] *University of Southampton, Southampton, United Kingdom*
[2] *Department of Applied Physics, KTH Royal Institute of Technology, Stockholm, Sweden*
[3] *Bristol Composites Institute, University of Bristol, Bristol, United Kingdom*

Corresponding author email: christopher.holmes@southampton.ac.uk

The digitalisation of engineered structures is being driven by the integration of advanced sensing and AI technology, enabling real-time monitoring, adaptive control, and autonomy of physical assets. Glass and photonic technologies, in particular optical fibre, are key in such development, as they offer robust sensing solutions that perform reliably in harsh environments, with minimal cabling intrusion.

This paper reports a recently developed class of silica-based optical fibers, termed high-aspect-ratio-flat-fiber (HARFF). This optical fibre type, examples of which are shown in Figure 1, has been developed to enable new optical sensing capabilities, supporting higher-fidelity and more ubiquitous sensing systems for digitalization[1,2]. This new family of optical fibres are manufactured using stack-and-draw combined with laser-aided techniques to mitigate surface-tension related deformation of the preform, which can be common in traditional flame-based methods.

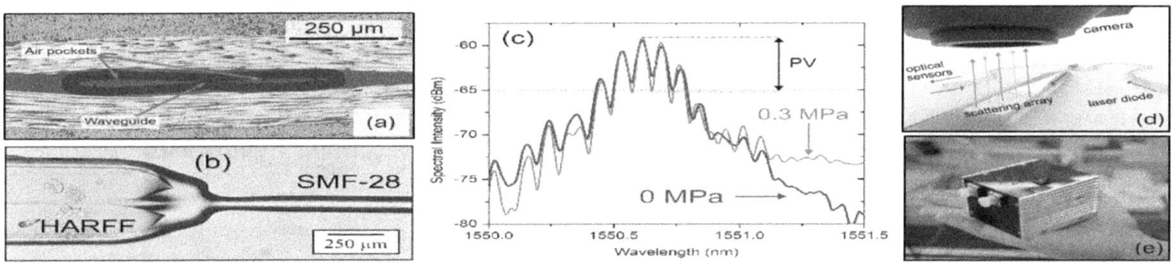

Figure 1 High aspect ratio flat fibre showing (a) cross section integration of dual-channel HARFF in carbon fibre composite (b) top-view of fusion splice to standard SMF-28, (c) a Fabry-Perot Fibre Bragg grating hydrostatic pressure response, (d)–(e) compact specklemeter based upon HARFF permitting compact and cost-effect optical fibre sensing

We shall report different designs of HARFF with tailored mechanical properties, suitable for enhanced in-process and through life monitoring of high-value composites[1] and spectral monitoring through specklemeter designs[2]. We shall report methods of fabrication using our unconventional, laser-aided methods, and their use to produce complex HARFF cross-sections. We will present the latest results in process monitoring, ranging from composite coupons to real-time systems deployed on autonomous aerial drones.

References

1. Holmes, C., Dulieu-Barton, J., Flexible photonics in carbon and glass fiber reinforced polymers for new multifunctionality: Exploring the advances, challenges, and opportunities Optical Materials: X, **20**(100277) (2023).
2. Falak, P., *et al.* Compact high-resolution FBG strain interrogator based on laser-written 3D scattering structure in flat optical fiber. *Sci Rep* **13**, 8805 (2023).

Phys. Chem. Glasses: Eur. J. Glass Sci. Technol. B, 2025, **66**(4), 180 DOI: 10.13036/17533562.66.4.AM68

Quantification of chemical replacement zones in glasses by means of ToF-SIMS, NRA and nanoindentation

K. Rein,[1] *A. Korn,*[2] *T. Schwarz-Selinger,*[3] *C. Volkert,*[2] *U. v. Toussaint*[3] *& Karl-Michael Weitzel*[1]

[1] *Department of Chemistry, Philipps-Universität Marburg, 35032 Marburg, Germany*
[2] *Institute for Material Physics, Georg-August-Universität Göttingen, 37077 Göttingen, Germany*
[3] *Max-Planck-Institute for Plasmaphysics (IPP), 85748 Garching, Germany*

Corresponding author email: weitzel@chemie.uni-marburg.de

Chemical replacement zones in solid state materials have relevance in many fields of material science. The most powerful approach for generating such replacement zones is the electric field assisted ion exchange (EFAIE) which induces uni-directional transport in solids and allows to manipulate the physical properties of solid electrolytes, especially near the sample surface.[(1)] *Native alkali cations such as Na^+ can, e.g. be exchanged by foreign cations introduced with, e.g. sputtered/molten electrodes.*[(1,2)] *The ion exchange can lead to, among other things, chemical hardening, improvement of antimicrobial properties or the formation of waveguides structures.*[(2–4)]

In this contribution we demonstrate the formation of pronounced replacement zones, and the quantification of the replacement zones by three different techniques. The alkali deuteron substitution (ADS) is a special form of the EFAIE, where Deuterium is converted to D^+ ions at a thin platinum electrode. An electric field applied drives the D^+ ions into the sample and induce replacement of the native (and mobile) alkali ions. This technique is an adaptation of the alkali proton substitution technique firstly introduced by Omata et al.[(5)] *We have implemented a modified version of this in our group recently.*[(6)]

Here, we present the results of an ADS experiment on an ion conducting D263T borosilicate glass, for which the native properties have been established before in a CAIT experiment.[(7)] *As a result of the ADS experiment, the native Na^+ and K^+ are replaced by D^+ ions up to a depth of 2 μm from the surface. The replacement zones are quantified by means of ToF-SIMS, nanoindentation and Nuclear Reaction Analysis (NRA). We compare the quality of the results of the different techniques. Ultimately, all three techniques arrive at the same characteristics of the replacement zone, however, with complementary information content including hardness and total charge carrier concentration.*

References

1. N. Takamure, A. Kondyurin and D. R. McKenzie, J. Appl. Phys. **125**, 175104 (2019).
2. M. E. Nordberg, E. L. Mochel, H. M. Garfinkel and J. S. Olcott, J. Am. Ceram. Soc. **47**, 215-219 (1964).
3. D. Guldiren, S. Aydin, Mater. Sci. Technol. C **78**, 826-832 (2017).
4. M. J. Li, S. Honkanen, W. J. Wang, R. Leonelli, J. Albert and S. I. Najafi, Appl. Phys. Lett. **58**, 2607 (1991).
5. T. Ishiyama, S. Suzuki, J. Nishii, T. Yamashita, H. Kawazoe and T. Omata, Solid State Ionics **262**, 856-859 (2014).
6. K. Rein, K.-M. Weitzel, J. Mat. Chem. A, **12**, 14117-14128, (2024)
7. J. Martin, S. Mehrwald, M. Schäfer, T. Kramer, C. Jooss, K.-M. Weitzel, Electrochimica Acta, **191**, 616–623 (2016)

Phys. Chem. Glasses: Eur. J. Glass Sci. Technol. B, 2025, **66**(4), 181 DOI: 10.13036/17533562.66.4.AM69

New MWIR (mid-wave infrared) fibre lasers for sensing

Thomas Readyhoof, Angela Seddon, Mark Farries & Sendy Phang

MIR Photonics Group, Faculty of Engineering, University Park Campus, University Park, Nottingham, NG7 2RD, UK

Corresponding author email: egytr1@nottingham.ac.uk

The MWIR (mid-wave infrared) spans the 2–10 mm wavelength range of the electromagnetic spectrum. The MWIR includes the atmospheric windows at 3–5 and 8–10 mm wavelength, with potential for atmospheric communications, direct chemical sensing, counter-measure applications in defence and collision-avoidance.

In 2021, we reported the world's first ever continuous wave MWIR fibre laser operating >4 mm wavelength. Such fibre lasers promise far higher beam quality than any other MWIR lasers viz.: OPOs (optical parametric oscillators); ICLs (interband cascade lasers); QCLs (quantum cascade lasers) and gas lasers like CO_2 and CO. In addition, fibre lasers exhibit large energy storage which uniquely enables high peak power pulses useful for generating MWIR wavelengths via nonlinear processes.

The PhD Programme builds on this first, MWIR >4 mm lasing fibre, extending the technology to include investigation of pulsed behaviour, further rare earth ion emitters across the MWIR and applying Bragg grating cavities, with demonstration of fibre laser sensing of chemical moieties of interest. The chalcogenide glass fibres, on which the brand new fibre lasers are based, are low loss, stable at ambient temperature, can be co-doped for potential tuneable emission across 2–10 mm wavelengths and pumped with low-cost near-infrared laser diodes. The chalcogenide glass low phonon energy and high refractive index means low non-radiative decay rates for efficient light emission and high cross-sections for short devices.

Phys. Chem. Glasses: Eur. J. Glass Sci. Technol. B, 2025, **66**(4), 182 DOI: 10.13036/17533562.66.4.AM70

Sculpting glass with light for future fibre optics

Calum A. Ross

Institute of Photonics and Quantum Sciences, Heriot-Watt University, Edinburgh, EH14 4AS, UK

Corresponding author email: calum.ross@hw.ac.uk

For 50 years, glass optical fibres have been a cornerstone of telecommunications networks, enabled by ultra-pristine silica glass which can be drawn to fibre with an attenuation of less than 0·2 dB/km. Recently, hollow-core fibres (HCFs), which guide light in air rather than glass, have succeeded the lowest-loss solid fibres for the first time, paving the way for a new era in fibre-optic communication. Offering lower latency and higher bandwidth than solid fibres, HCFs are experiencing a surge recently, driven by developments in AI and cloud networking. Despite this surge, HCFs are limited in design by manufacturing challenges associated with the conventional "stack-and-draw" method, whereby glass rods and capillaries are stacked longitudinally to form the fibre preform.

To unlock the full potential of HCFs, we are exploring radically new approaches to manufacturing fibre preforms based on advanced laser processing. In this approach, arbitrarily structured silica discs are laser fabricated and bonded axially to form the preform, which is then drawn to optical fibre using a conventional fibre drawing tower. We hope that this method will enable the manufacture of freeform fibres with unconstrained core-cladding structure.

In my talk, I will give an introduction to ultrafast laser processing of glass and discuss some of the unique capabilities it offers. I will then present our work on the development of freeform optical fibres, and discuss aspects of glass processing including laser processing, chemical etching, glass bonding and fibre drawing.

Phys. Chem. Glasses: Eur. J. Glass Sci. Technol. B, 2025, **66**(4), 183 DOI: 10.13036/17533562.66.4.AM71

Manifestation of the mixed alkali effect on viscosity in the $K_2O/Na_2O–CaO–SiO_2$ system

Wei Deng & Paul A. Bingham

College of Business, Technology and Engineering, Sheffield Hallam University, Sheffield S1 1WB, UK

Corresponding author email: wei.deng@shu.ac.uk

Leveraging the opportunity of introducing biomass ash as the raw materials for silicate glass, this study aims to revisit the mixed alkali effect phenomenon in ordinary silicate glass through the apparent parameter of viscosity. Glasses with nominal molar composition (13·626–n)Na_2O–nK_2O–13·747CaO–72·626SiO_2 (mol%) where n=0, 2·5, 5, 7·5, 10 and 13·626 were prepared and characterised through DSC, thermal expansion and high-temperature viscosity. At lower temperatures, the mixed alkali effect was observed through changes in T_g. By fitting the glass viscosity to the modified MYEGA equation and comparing it with the theoretically calculated mechanically mixed viscosity, we observed, defined, and quantitatively described the nonlinear viscosity behaviour influenced by the mixed alkali effect at high temperatures. Despite differences in the systems, the nonlinear viscosity behavior is universal, further advancing our understanding of the mixed alkali effect and its boundaries.

Phys. Chem. Glasses: Eur. J. Glass Sci. Technol. B, 2025, **66**(4), 184 DOI: 10.13036/17533562.66.4.AM72

Tunable luminescence and realisation of warm white light via energy transfer from Dy^{3+}/Er^{3+}/Sm^{3+} triply doped glasses

Adon Jose[1] *&* *P. R. Biju*[2]

[1] *St. Stephen's College Uzhavoor, Kerala 686634, India*
[2] *School of Pure & Applied Physics, Mahatma Gandhi University, Kottayam, Kerala 686560, India*

Corresponding author email: prb.mgu@gmail.com

The spectroscopic properties of $BaO–ZnF_2–K_2O–SiO_2–B_2O_3$ glasses doped with Dy^{3+}, Er^{3+} and Sm^{3+} – individually, in pairs, and in combination – were investigated using photoluminescence excitation and emission spectra, decay profiles, CIE color coordinates, and correlated colour temperature (CCT) values. The fluorescence spectra of triply doped Dy^{3+}/Er^{3+}/Sm^{3+} borosilicate glasses under various excitation wavelengths (348, 374, 379 and 400 nm) revealed the simultaneous presence of Dy^{3+}, Er^{3+}, and Sm^{3+} emission bands, highlighting active energy transfer processes, which were analysed in detail. A notable decrease in the luminescence intensity of Er^{3+} and Sm^{3+} bands was observed with increasing Dy^{3+} concentration, while Dy^{3+} emission bands exhibited a consistent enhancement. The luminescent kinetics of the samples were thoroughly examined through recorded decay profiles, with a decrease in lifetime values confirming the proposed Er^{3+} (or Sm^{3+}) to Dy^{3+} energy transfer pathways. Energy transfer efficiency was found to increase significantly with higher Dy^{3+} ion concentration and was systematically interpreted using partial energy level diagrams. The estimated CCT values and CIE colour coordinates demonstrated a transition from reddish-orange to warm white light, underscoring the material's potential for lighting applications.

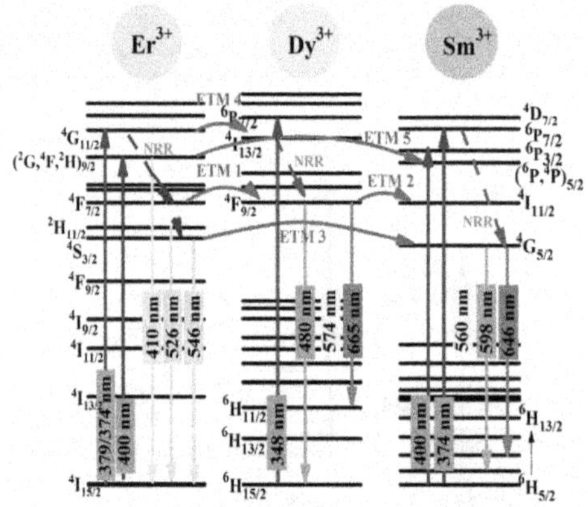

References

1. A. Jose et.al. *J.Lumin* 236 (2021) 118105.

Phys. Chem. Glasses: Eur. J. Glass Sci. Technol. B, 2025, **66**(4), 185 DOI: 10.13036/17533562.66.4.AM73

Eu^{3+}/Dy^{3+} co-activated multicomponent borosilicate matrix: multifunctional glass for lighting and solar energy harvesting

Adon Jose[1] *& P. R. Biju*[2]

[1] *St. Stephen's College Uzhavoor, Kerala 686634, India*
[2] *School of Pure & Applied Physics, Mahatma Gandhi University, Kottayam, Kerala 686560, India*

Corresponding author email: prb.mgu@gmail.com

Rare-earth-doped luminescent glasses, particularly lanthanide-activated materials, have garnered significant interest in modern luminescence research. This study investigates the luminescence features of Dy^{3+}/Eu^{3+} co-activated multicomponent borosilicate glasses ($10BaO–10ZnF_2–10K_2O–20SiO_2–(49−y)B_2O_3–1Eu_2O_3–yDy_2O_3$, y=0·1, 0·25, 0·5, 0·75, 1 mol%), synthesized via the melt-quenching method. XRD analysis confirms the amorphous nature of the glasses. A comprehensive luminescence study – including photoluminescence excitation, emission, decay profiles, and colorimetric analysis – was conducted to explore energy transfer mechanisms. The spectral overlap between Eu^{3+} excitation (λ_{em}=612 nm) and Dy^{3+} emission (λ_{exc}=348 nm) bands in singly doped glasses suggests the presence of energy transfer in the co-doped multicomponent borosilicate glass. The increasing intensity of the $^5D_0\rightarrow{}^7F_2$ (Eu^{3+}) emission under 348 and 392 nm excitation, correlating with Dy^{3+} concentration in Eu^{3+}/Dy^{3+} co-activated multicomponent borosilicate glass, validates the predicted energy transfer pathways observed in the spectral overlap diagram. Decay profile analysis indicates a 42% energy transfer efficiency between the rare-earth ions in the prepared matrix. Additionally, a comparison of solar irradiance spectra with the photoluminescence characteristics of the prepared glasses was conducted to evaluate their potential as down-conversion materials for solar cells. The strong white and red emission characteristics of these glasses make them promising candidates for solid-state lighting and solar energy harvesting applications.

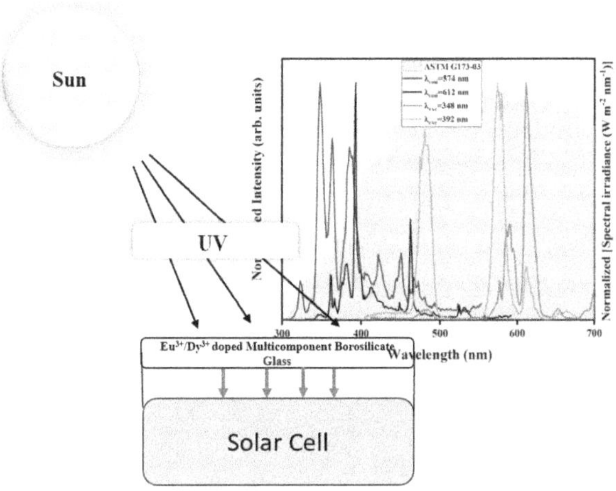

References

1. L. Yuliantini et.al. *Opt Mater (Amst)* 149 (2024) 115070.
2. T.S. Dhapodkar et.al, *Mater Today Chem* 24 (2022) 100938.

Phys. Chem. Glasses: Eur. J. Glass Sci. Technol. B, 2025, **66**(4), 186 DOI: 10.13036/17533562.66.4.AM74

Multicomponent glasses for hollow core antiresonant fibres

Mahmudur Rahman, Nicholas White, Ghafour Amouzad Mahdiraji, Gregory Jasion & Francesco Poletti

Optoelectronics Research Centre, University of Southampton, Southampton, SO17 1BJ, UK

Corresponding author email: m.rahman@soton.ac.uk

Hollow core fibres (HCFs) have emerged as a promising platform for ultralow-loss light guidance, overcoming the intrinsic Rayleigh scattering loss of silica by confining light predominantly in a central air core. Recent advances in F300 fused silica HCFs have demonstrated even lower propagation loss than highly pure silica-based fibre in the visible and near-infrared (Vis-NIR) regions, but these still suffer attenuation from optical scattering caused by the intrinsic glass roughness of the surfaces surrounding the core. The magnitude of this surface roughness can, however, be reduced by operating at a lower glass transition temperature (T_g). In this study, we experimentally investigate the feasibility of fabricating HCFs from multicomponent glasses with a T_g lower than fused silica. Due to the low percentage of the optical field propagating in the glass (90 PPM), a glass with a relatively high bulk material absorption (200 dB/km) could still be a feasible candidate for producing HCFs with comparable attenuation.[1] Using a combination of spectroscopic analysis and literature data, we compare the absorption profiles of several commercially available SCHOTT glass rods and evaluate their suitability for HCF fabrication.

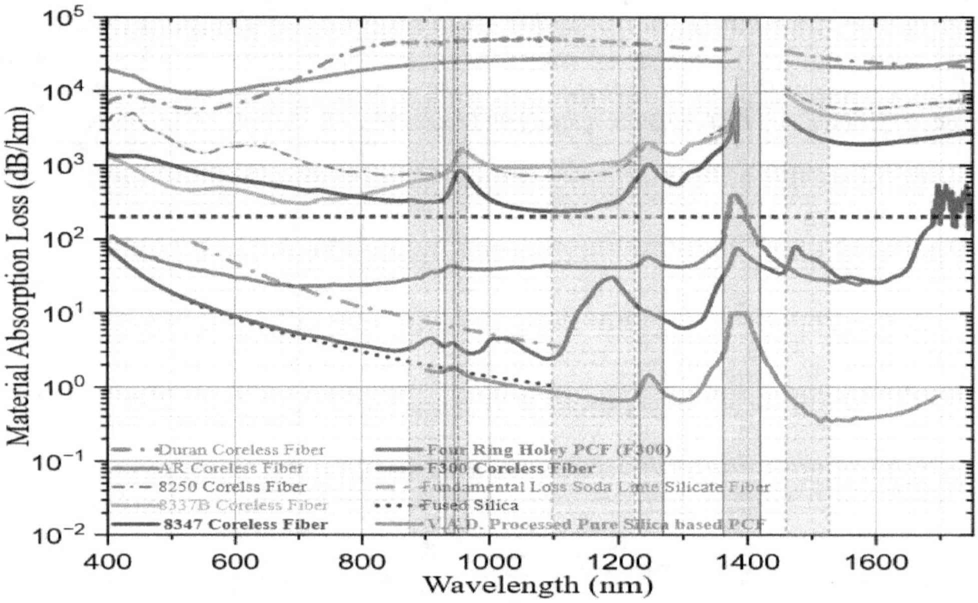

Reference

1. Numkam Fokoua, Eric, et al. "Loss in hollow-core optical fibers: mechanisms, scaling rules, and limits." *Advances in Optics and Photonics* 15.1 (2023): 1-85.

Phys. Chem. Glasses: Eur. J. Glass Sci. Technol. B, 2025, **66**(4), 187 DOI: 10.13036/17533562.66.4.AM75

The evolving future of sustainable glass art practice

Jessamy Kelly

Senior Lecturer in Glass, Edinburgh College of Art, University of Edinburgh

Corresponding author email: jessamy.kelly@ed.ac.uk

In the ever-evolving landscape of our times, climate change is a pressing issue that sits prominently on the agenda of many glass art practitioners. It is crucial for us as glassmakers to critically reassess and innovate our artistic glass practices. In our post-industrial era, marked by profound instability and uncertainty, the concept of the Anthropocene highlights how human activity has drastically and irreversibly altered our climate and environment. Glassmaking, traditionally seen as an industrial activity, involves extracting and exploiting natural resources and materials, raising serious concerns about the impact of this discipline on our environment.

This issues a serious call for glass makers to prioritise sustainable practices and chart a progressive future for their craft. As responsible practitioners, it is crucial for them to develop and adopt sustainable methods that minimize or reduce the environmental impact of their work. Glass is notable for its sustainable attributes as a material that is infinitely recyclable when properly cleaned and sorted. Using recycled glass not only offers sustainable solutions but also plays an active role in reducing environmental harm. A shift towards sustainable and locally sourced materials is increasingly evident among glassmakers, many of whom focus on the principles of reducing, reusing, and recycling glass in their practices. However, the sector still faces significant challenges that require additional attention. The processing of glass at high temperatures consumes substantial energy, and glass production often involves considerable water wastage during cold processing stages. Addressing these issues is imperative for the future sustainability of glassmaking.

In my own creative practice, I am motivated by environmental concerns and make a conscious decision to use recycled glass, incorporating reuse in my glass making process, and seeking lower-impact alternatives for glass processing. Through this, I aim to address broader environmental issues such as material extraction, deforestation, and natural resource exploitation. This paper will explore how new works addressing these concerns have been created. This paper will be extended by examining a range of case studies of glass makers who are also reassessing the impact of their work and adopting responsible production and consumption practices. Moreover, this paper will conclude with a discussion on how artistic expression can elevate voices and catalyse meaningful shifts in the ways we approach and relate to glassmaking. It will underscore the pivotal role of glass as a medium through which makers can express environmental concerns and demonstrate that craft can be a potent form of activism, especially when advocating for climate action.

Keywords – Art glass, sustainability, glassmaking, recycled glass, low-impact, environmental

Phys. Chem. Glasses: Eur. J. Glass Sci. Technol. B, 2025, **66**(4), 188 DOI: 10.13036/17533562.66.4.AM76

Re-contextualising scientific glass: unlocking alternative perceptions of value in high jewellery

Estelle Burton

(Jewellery) Design MA, Central Saint Martins, University of the Arts, London, UK

Corresponding author email: estellemburton@gmail.com

This talk explores the cultural and material potential of precision-engineered scientific glass within the realm of high jewellery, positioning it as a catalyst for rethinking value hierarchies and encouraging cross-disciplinary collaboration in craft. Drawing on 18 months of embedded field research within a range of UK industrial glass manufacturers, the study investigates how production cultures operate, the sharing of tacit knowledge, and the parallels between high-volume production in scientific glassmaking and the bespoke high jewellery atelier.

The research highlights opportunities for knowledge exchange, shared tooling, and the creative reuse of non-recyclable waste, all examined through the lens of Luminaire, a capsule jewellery collection developed as a proof of concept. The lecture will introduce the process of reframing rare industrial artefacts by combining elements of glasswork with traditional jewellery-making materials and techniques.

Each jewel in the collection transforms precision-engineered conducting and insulating components into a central design feature. Commonly used in scientific instruments, medical devices, and aerospace technology, these elements employ techniques specifically designed to fuse glass with specialist metal alloys. This allows electrical signals or light to pass in a manner that is completely airtight and resistant to heat, pressure, and corrosion. With a robustness and unique industrial beauty, this research endeavours to present glass in a new light, ideally positioned to transform perceptions of how it may be worn on the body.

Both disciplines demand years of training and a complete commitment to quality. They share similar tools, processes, and monetary value; however, the materials used and public perceptions differ sharply. High jewellery enjoys prestige and visibility, whereas scientific glass artefacts-although optically striking and intriguing jewels in their own right-are primarily valued for their technical function and remain hidden inside extraordinary machines and equipment, with their beauty unseen by most of the population.

By combining these components with precious materials and jewellery-making techniques, the collection explores both their aesthetic and structural potential. This approach opens up new possibilities for sustainable practice through cross-disciplinary collaboration between science and craft, offering the potential to renegotiate notions of preciousness, safeguard endangered making practices, and present a new model for sustainable, cross-disciplinary craft clusters.

* Estelle Burton Contemporary Jewellery, Editor Board *Jewellery History Today*, Lecturer University Centre Hastings and recipient of the Oldfield Award 2024.

Phys. Chem. Glasses: Eur. J. Glass Sci. Technol. B, 2025, **66**(4), 189 DOI: 10.13036/17533562.66.4.AM77

Yellow colour of stained-glass windows: a French historical perspective

Déa Jaïs,[1] Théo Caroff,[1] Laurence Galoisy,[1] Claudine Loisel,[2] Barbara Trichereau[2] & Georges Calas[1]

[1] *IMPMC Institut de Minéralogie, de Physique des Matériaux et de Cosmochimie, Sorbonne Université, Paris, France*
[2] *LRMH Centre de recherche sur la conservation, CNRS, Laboratoire de Recherche des Monuments Historiques, Champs-sur-Marne, France*

Corresponding author email: georges.calas@upmc.fr

Yellow colour has been and still is widely used in stained glass art, produced by several colouring processes, efficient enough to overcome the optical absorption constraints imposed by the limited thickness of window glasses. This was achieved by using processes able to produce intense yellow shades, as charge transfer in amber chromophore S^{2-} to Fe^{3+}, plasmon resonance of metallic silver nanoparticles, or light absorption by semiconducting nanoparticles of CdS-Se. This study compares the yellow colour of French stained glasses from major French monuments, which have been dated from 14th to 21st centuries. This review encompasses a medieval Transfiguration panel from Troyes (circa 1170–1180); Renaissance stained glass windows from Saint-Etienne-du-Mont Church (Paris), circa 1612–1622; "Art Nouveau" glass tiles by René Lalique; from Saint-Nicaise Church (Reims) 1928; "contemporaneous stained glass windows by Henri Matisse, from the Rosary Chapel, Vence, 1951; contemporaneous stained glass windows by Antoine Le Bihan and Claire Babet, from Saint-Gatien Cathedral, Tours, 2024. Special attention was given to this fail among original and replacement pieces. The chemical composition of these glasses was obtained by nondestructive techniques such as PIXE/PIGE on the New AGLAE accelerator in the Louvres Museum in Paris. Optical absorption spectra in the UV-visible-NIR and Raman spectra were obtained by contactless techniques with portable instruments. These analyses show that amber chromophores were first used for obtaining yellow glasses at the end of the 12th century. Silver yellow appeared in the early 14th century, followed by cadmium sulfo-selenide nanoparticles ("quantum dots") in the early 20th century. Today, these chromophores are still used in contemporary stained glass, offering an extremely rich and sought-after palette of yellow glass shades. Colorimetry modeling indicates that the yellow shade caused by these pigments is actually slightly different depending on the chromophore used. As a matter of fact, all these chromophores are still in use in contemporary stained glass, offering an extremely rich and sought-after palette of yellow glass shades.

Phys. Chem. Glasses: Eur. J. Glass Sci. Technol. B, 2025, **66**(4), 190 DOI: 10.13036/17533562.66.4.AM88

"The Salts of the Rare Metals": a physical-chemical description of Frederick Carder's aurene glass

Annika Blake-Howland & Doris Möncke*

[1] *New York State College of Ceramics at Alfred University, 1 Saxon Drive, Alfred, New York, 14802, USA*

Corresponding author email: AB58@alfred.edu

Frederick Carder, cofounder of Steuben Glass in Corning, New York, USA, filed a trademark for a gold-colored iridescent glass he named Aurene on 6 September 1904. Carder was inspired by archaeological Ancient Roman glass that had iridescent surfaces due to being in a wet-cycling burial environment. Throughout the development and manufacturing of Aurene Carder kept notes, frequently in code, in pocket notebooks. From these notebooks we can determine the glass is approximately 49 mol% SiO_2 and 31 mol% PbO.[(1)] The remainder of the melt includes B_2O_3, Na_2O, K_2O, NiO and Ag_2O.[(1)] The general production process of an Aurene objects, as deduced from the heavily coded notebooks, was to start with a melt with the components listed previously. The object was then blown under reducing conditions, creating an exterior layer of metal oxides in their reduced states[2]. Next, the object was sprayed with iron chloride followed with tin chloride and heated in an oxidizing flame to produce the iridescent effect that results from thin layers of varying refractive index, and varying incident angle as provided by the round curves of vessels.[(2)] This complex process, done entirely by hand by expert gaffers, created translucently walled iridescent gold blown glass decorative arts objects.

Despite the art historical importance of Carder's Aurene glass and the scientific study of other late nineteenth and early twentieth century iridescent glass artists, Aurene glass has not been studied scientifically. However, the iridescent gold appearance of the Aurene is of art historical interest as it was produced during a time of great change in glass decorative arts in the United States. Similarly, it is of interest to art conservators due to an informal understanding that Aurene glass is less prone to surface damage than other similar contemporary glasses. The nickel, silver, tin, and iron ions all contribute to the final appearance of the Aurene glass. Thus, the primary focus of this presentation is on the distribution of metal ions within the surface of the artefacts, rather than in the bulk, and how agglomeration of these highly polarizable ions changes the appearance of the surface. Through the use of spectroscopic analysis, the Aurene bulk glass and the iridescent surface are studied, tracking changes throughout the Aurene glass vessel wall cross section and focusing on the distribution of metallic ions within the glass. Additionally, the difficulties in studying cultural heritage samples will be discussed, including sampling constraints and destructive testing methods.

References

1. Carder, F. Red Pocket Sized Notebook with Glass Recipes. https://cmog.primo.exlibrisgroup.com/ permalink/01CORNING_INST/1531shl/alma991121823504126 (accessed 2023-10-13).
2. Gardner, P. V. *The Glass of Frederick Carder*, 4th Printing.; Crown Publishers, Inc.: New York, 1971.

Phys. Chem. Glasses: Eur. J. Glass Sci. Technol. B, 2025, **66**(4), 191 DOI: 10.13036/17533562.66.4.AM89

The charm of combining glass with other materials in artistic creation

Jianyong Guo

Shandong University of Art & Design, China

Corresponding author email: gjy36900@163.com

In contemporary art creation, the materials and languages used by artists, as well as the concepts and themes they intend to express, have all undergone significant changes. Glass is a material that is both traditional and contemporary. With the continuous maturation and enrichment of glass-making techniques, artists' pursuit of the beauty of materials has become more diversified. The combination of glass with different materials has, to a large extent, broadened the development space of glass art creation, enhanced the visual effect and spatial perception of glass art works, and has certain theoretical and practical significance. Glass is highly malleable, while metals, ceramics, fibers, woods and stones all have different textures, patterns and colours. By means of glass processing methods such as casting, slumping, flame working, inlaying and bonding, these materials are reasonably combined with glass. By taking advantage of the complementary physical and chemical properties of different materials, the performance limitations of a single material can be broken through. This helps to promote the innovation of design concepts and technical processes, achieve cross-border dialogue in material language, and enrich the expression forms of glass works. It can also carry the artist's emotions and thoughts more perfectly, bringing a novel visual experience to the viewers (see figure 1). This paper analyses and expounds the application methods and significance of comprehensive materials in glass art creation, hoping to bring some inspiration to other art creators.

Figure 1. Jianyong Guo, Dancing, glass, copper, metal oxide powder, 30×50×20 cm, 2023.

Phys. Chem. Glasses: Eur. J. Glass Sci. Technol. B, 2025, **66**(4), 192 DOI: 10.13036/17533562.66.4.AM80

Glass & Light: materialities & metaphors

Charlie Murphy

Corresponding author email: charlie@charliemurphy.co.uk

Artist Charlie Murphy develops a wide range of art installations and public engagement strategies which utilise the sculptural, optical and metaphoric dimensions of glass.

Working in collaboration with new technologist Robin Bussell, Charlie is developing new projects which reform scientific glass to visualise insights into human anatomy, health and scientific discovery.

Introducing the scientific research, creative responses and making processes which inform her work, Charlie will showcase projects in development with Birmingham (Centre for Custom Medical Devices), Loughborough & Aston Universities (National Centre for Neuromorphic Computing).

Short biog: Charlie Murphy has established a vibrant, interdisciplinary practice spanning dynamic sculptural installations, digital media, performance and live events for many arts, science, public health, museum and community contexts.

She create dramatic sculptural installations, performances and events which explore material questions about our bodies, health & relationships. Widely known for her popular touring 'kiss-in'events, art & science collaborations and performative events, her projects often involve participation and collaboration of some kind.

Websites: www.charliemurphy.co.uk www.brainsinadish.org

Instagram: charliemurphy_artiststudio

Phys. Chem. Glasses: Eur. J. Glass Sci. Technol. B, 2025, **66**(4), 193 DOI: 10.13036/17533562.66.4.AM81

Prototyping creative circularity in luxury tableware

Lulu Harrison,[1,2] *Paul Bingham,*[3] *Sabrin Samad,*[3] *Tony Bell,*[3] *Ben Gough,*[4] *& Elliot Walker*[5]

[1] *Lulu Harrison Studio, Oxfordshire, UK*
[2] *Maison/0, Central Saint Martins, Granary Building, 1 Granary Square, London N1C 4AA, UK*
[3] *Materials and Engineering Research Institute, Sheffield Hallam University, Sheffield, S1 1WB, UK*
[4] *The Glass Hub, Stowford Manor Farm, Trowbridge BA14 9LH, UK*
[5] *Blowfish Glass, Unit 15, Red House Glass Cone, Stourbridge DY8 4AZ, UK*

Corresponding author email: info@luluharrisonstudio.com

Glassmaking is traditionally energy-intensive and relies on finite raw materials. This project explores creative circularity in luxury tableware by sourcing waste materials from local industries to develop a sustainable glassmaking process promoting circular economy. The aim is to reduce the use of non-renewable materials to reduce energy consumption, while investigating how ancient glassmaking methods could inspire a new approach to luxury. Herein, a regional waste material sourcing strategy was adopted that would contribute to lowering the melting temperature of the glass batch while reducing carbon emissions in the glassmaking process. Through establishing a local network of stakeholders for material sourcing and fabrication, the final glass recipe included wine lees, biomass ash, quagga mussel shells, local sand, waste cullet and recycled glass. After thorough characterisation of these raw materials via x-ray fluorescence (XRF) and x-ray diffraction (XRD), a recipe was optimised after 22 glass melting trials, celebrating naturally occurring colour from waste raw materials. This process was then upscaled to be melted in a glassblowing studio, 'The Glass Hub' using a nano-melt furnace (Images 1,2).

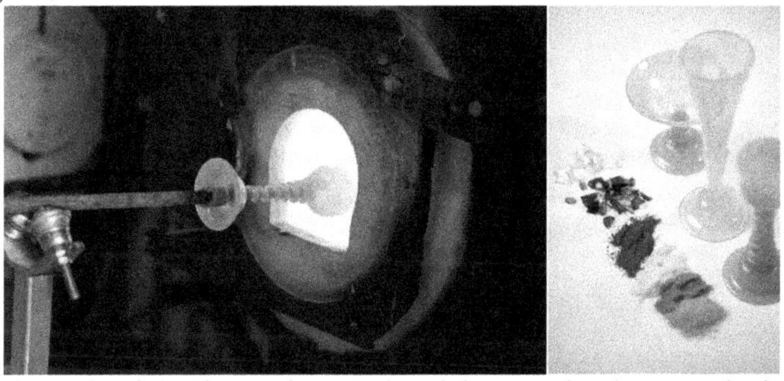

The project then progressed to 'Blowfish Glass' where 90 kg of the same batch was melted in a tank furnace, creating bubble-free, hand-blown glass collections, each with their own playful theme (Images 3–5).

This work demonstrates that by combining art & science / industry & academia, there could be an exciting future for glassmaking using local, abundant raw materials and waste by-products to create experimental and sustainable glass batches.

Phys. Chem. Glasses: Eur. J. Glass Sci. Technol. B, 2025, **66**(4), 194 DOI: 10.13036/17533562.66.4.AM82

Cycles of transformation: exploring lifecycles in glass art making and material reuse

Gregory Alliss

Edinburgh College of Art, The University of Edinburgh, 74 Lauriston Place, Edinburgh, EH3 9DF, UK

Corresponding author email: S1795601@sms.ed.ac.uk

Glass art making occupies a unique position at the intersection of material science and creative expression. The paper explores the lifecycles of glass within the context of contemporary glass art making, from sourcing and production to artistic transformation, reuse, and material processing streams.

The lifecycles presented draw on case studies from studio glass artists, archival material, and interviews with practitioners who are redefining glass use in their practice. By examining these lifecycles through artistic lens, we consider not only the physical changes the material undergoes, but also the evolving narratives it carries through each phase.

The paper addresses how artists engage with recycled studio glass, post-consumer waste glass, and found objects, and how these choices influence both form and meaning in their work.

The lifecycles presented are set within the framework of wider lifecycles of the glass industry. It is acknowledged that glass used and reused within glass art making is a very small percentage of total glass used globally.

By framing glass art as lifecycle process rather than a linear act of creation, this research invites artists, curators and glass manufacturers, to rethink material agency and the role of craft in shaping sustainable artistic futures.

Keywords: Glass Art, Material Lifecycle, Sustainability, Craft Practices, Recycled Materials, Contemporary Art.

Phys. Chem. Glasses: Eur. J. Glass Sci. Technol. B, 2025, **66**(4), 195 DOI: 10.13036/17533562.66.4.AM83

Recovering secrets from historic glass-waste

Colin Brain

Corresponding author email: cjsm132@gmail.com

Excavated glass-making-waste residues have the potential to help us better understand how glass was made in the past. This paper centres on one such group found at the Museum of London Archaeology (MOLA) excavation site WEO19 in Weston Street, Southwark. Initial appraisal of these finds suggested that they were indeed rare glass-manufacturing waste probably dating from the 1670s. Unusually this means that they are historic, in the sense that there is period documentation available to help in the interpretation of the scientific analysis results. Some of this mentions 'glass secrets'. In the recent past it has been taken for granted that these secrets concerned the 'invention' of novel raw materials, particularly involving George Ravenscroft and 'lead' glass.

Thus this waste material has the potential to provide a better understanding of glass-making at this critical stage in the development of the British glass industry and perhaps to allow us tease-out some additional 'secrets'. The key approach reported here was to estimate the viscosity temperature relations for selected fragments based on their analysed elemental compositions. This uses the work pioneered by Lakatos et al in the 1970s. Although this technique has been used before, it rarely features in glass history and heritage publications. In this case it showed that the WEQ19 glass analysed had surprisingly little dispersion (typically ±1%) in the estimated melting temperatures, consistent with these being the output of a single furnace type and, beyond that, there seemed to be only three melting temperatures in use across all the similarly-dated sites investigated (1160, 1210 & 1270°C).

Phys. Chem. Glasses: Eur. J. Glass Sci. Technol. B, 2025, **66**(4), 196 DOI: 10.13036/17533562.66.4.AM84

Reproducing the colours of the Lycurgus Cup

Lisa Hollands,[1] *Caroline Jackson*[2] *& John Parker*[1]

[1] *School of Chemical, Materials and Biological Engineering, The University of Sheffield, Sir Robert Hadfield Building, Mappin Street, Sheffield S1 3DU, UK*
[2] *School of History, Philosophy and Digital Humanities, University of Sheffield, Minalloy House, 10–16 Regent Street, Sheffield, S1 3NJ, UK*

Corresponding author email: l.hollands@sheffield.ac.uk

Recently, we were challenged to recreate a dichroic glass with similar optical characteristics to the Lycurgus Cup for an exhibition celebrating light as part of Science Week. The Lycurgus Cup, which is currently housed at the British Museum, is a unique example of an almost intact Fourth century Roman cage cup which has the characteristics of appearing opaque jade in reflected light, but red in transmitted light. PPM levels of nanoparticles containing gold and silver have previously been identified as being responsible for the dichroic effect of the glass, however attempts at reproducing the glass itself have not been well publicised. Over an initial 4 month period, we used information gathered from previous studies of the cup and an understanding of Roman glass manufacturing as a starting point. Altering the batch materials to attempt to mimic Roman starting materials and modifying melting conditions used, we produced a glass with an effect similar to the cup the day before the exhibition opened. A subsequent programme of work has enabled us to continue melt trials and start to characterise them to compare our findings to those of this historic object.

Phys. Chem. Glasses: Eur. J. Glass Sci. Technol. B, 2025, **66**(4), 197 DOI: 10.13036/17533562.66.4.AM85

Is glass composition a unique finger print for a 19th century glass maker?

J. M. Parker

Turner Museum of Glass, Department of Materials Science & Engineering, University of Sheffield, Portobello St, Sheffield S1 4DT, UK

Corresponding author email: j.m.parker@sheffield.ac.uk

The background behind the question in the title of this presentation originated from a glass collector who wished to link an article created by a particular glass artist to the specific manufacturer of the blank he used. Also archaeologists have many years been able to categorise the compositions of various ancient glasses to particular regions and periods in history based on compositional analysis, in particular using XRF and isotopic analysis by mass spectrometry.

At about the same time the speaker received a copy of a recipe book from a 19th century glass maker which lists batch compositions on an almost daily basis. The talk will examine the possibility of using various non-destructive methods to link the finished product to a particular manufacturer. In particular he will consider the accuracy of methods available and the likely variability of the results.

Unlocking the mysteries of glass

One day course: Wednesday, 3 September, Cambridge, UK
Time: Part I: 09.00–13.00; Part II: 14.00 –17.00.

Who can benefit:

Recent professionals in glass related industries with little or no formal background in glass. Students (undergraduate or graduate) and post-doctoral fellows. More seasoned professionals for the fun of it and to make lively interaction.

Outcome:

At the end of the course, you should develop an understanding of commercial glass compositions, glass structures, why glass often behaves mysteriously in the way it does, and which properties and attributes benefit certain commercial applications. (Participation in the course requires additional registration on the conference website.)

Part I Fundamentals of Glass Science & Technology

Topics covered: Families of commercial glasses; What is glass? Glass structures; Metastable phase separation; Viscosity; Glass transition range behaviour; Chemical durability; Hardness, Strength and Strengthening; Q&A

Instructor: Professor Arun Varshneya, Alfred University, USA

About the Instructor: Professor Varshneya is nicknamed *The Glass Guru* and has taught almost all undergraduate and graduate level courses in glass science/technology at Alfred University over his 30 years tenure. He is the immediate past president of the Society of Glass Technology, an Honorary FSGT, a Distinguished Life Member of the American Ceramic Society and a recipient of the ICG President's Award. Varshneya is the president of Saxon Glass Technologies, Inc of Alfred and is recognized for the chemically strengthened glass cartridge in EpiPen autoinjectors.

Course notes:

Key figures will be available as a handout. The course generally follows the textbook, *Fundamentals of Inorganic Glasses* third edition by Arun Varshneya & John Mauro (Elsevier). Students are encouraged to buy a copy from Amazon and bring to the class (for autograph!).

Part II: Fractographic analysis of glass products

Topics covered: Basics of glass fractographic analysis. Hands on examination of glass fractures.

Instructor: Daniel Capon, Technical Manager, Glass Technology Services (Sheffield).

About the Instructor: In his role as Technical Services Manager at Glass Technology Services, Daniel is responsible for overseeing this area of the business covering the extensive testing facilities and laboratories at GTS including chemical analysis, mechanical testing and materials analysis such as XRF. Daniel has a BEng in Materials Engineering and over 25 years' experience working with glass. He is a specialist in glass analysis and an expert in XRF and SEM work in particular. He is experienced in both the analysis of and remedial actions for hot glass defects such as stones, cord and bubbles and is an active member of the International Commission on Glass (ICG) working with Technical Committees, including Chairman of TC02 and was a contributor to the development of key studies relating to delamination of glass vials and drug interaction. He regularly provides training to a range of different international organisations working with glass manufacturers, pharmaceutical companies and beverage brands, regularly speaking on webinars and at events.

Learn about glass strength, damage and an analytical approach to failure investigation with one of the leading experts in glass. The session will cover the basics of glass fractography and be interactive, allowing for questions and discussion.

This technical session is suitable for students and people new to the glass industry to experienced glass technologists, quality assurance and product complaints professionals and features a combination of technical presentation, practical exercises, case studies and practical investigations to help delegates understand glass strength, weaknesses, failure and interpretation of evidence to determine causation.

What you will learn about: introduction to glass strength; fracture markings; breakage origins; breakage patterns; impact failures; internal pressure failures; thermal failures; hydrodynamic failures; vertical load failures.

Unlocking the Mysteries of Glass @£260.00 or book the conference for Monday and Tuesday the course on Wednesday for £660

CONFERENCE DIARY

7–11 September
34th Annual Conference of the European Society for Biomaterials, Turin, Italy.
Organizing Secretariat, Via Roma 10, 16121 Genova, Italy. Email esb2025@aristea.com Web esb2025.org

8–12 September
23rd Congress of the AIHV, Mainz, Germany.
Email aihvsecretary@gmail.com Web aihv.org/

10–12 September
Glasspex/Glasspro India 2025, Mumbai, India.
Messe Düsseldorf India, 20th Floor, Tower A, Building No. 5, DLF Cyber City, Phase II, Gurugram - 122002, India. Email info@md-india.com Web glassproindia.com

11 September
AI and Energy Conservation – The Future of Sustainability, the 15th AIGMF conference, Mumbai, India.
Email info@aigmf.com Web aigmf.com/future-events.php#11

13–21 September
Venice Glass Week, Venice, Italy.
Email info@theveniceglassweek.com Web www.theveniceglassweek.com

15–19 September
Sumglass 2025, Wuhan, China.
Dr Lifeng Li, Conference Secretary. Email lifeng.li@whut.edu.cn Web sumglass2025.scimeeting.cn/en/web/index/25180

16–16 September
44th Annual Cement & Concrete Conference, Leeds, UK.
The Institute of Materials, Minerals & Mining on behalf of the Cementitious Materials Subgroup, 297 Euston Road, London NW1 3AD, UK. Web www.iom3.org

16–19 September
Vitrum 2025, Milan, Italy.
17 September
Decarbonisation in the glass industry, Milan, Italy.
Email vitrum@vitrum-milano.it Web www.vitrumlife.it

28 September–1 October
MS&T25, Materials Science and Technology exhibition and conference incorporating ACerS 127th Annual Meeting, Columbus, Ohio, USA.
The American Ceramic Society, 550 Polaris Pkwy, Ste 510, Westerville, OH 43082, USA. Email customerservice@ceramics.org Web www.matscitech.org

3–4 October
ICON Stained Glass Group Annual Conference 2025, Canterbury, UK.
The Institute of Conservation, 106-109 Saffron Hill, London EC1N 8QS, UK. Web www.icon.org.uk

6–9 October
86th Glass Problems Conference and Symposium, Toledo, Ohio, USA.
Glass Manufacturing Industry Council, 550 Polaris Parkway, Suite 510, Westerville, OH 43082-7045, USA. Web gmic.org/glass-problems-conference/

7–9 October
World Soda Ash Conference 2025, Palma de Mallorca, Spain.
David Coates, Executive Director – Global Conferences and Events. Email dcoates@opisnet.com Web chemicalmarketanalytics.com/events/world-soda-ash-conference/

10–12 October
2025 Glass Olympics and Symposium, Alfred, NY, USA.
Chrissy Ziese, Special Programs Coordinator for the Arts, Alfred University, USA. Email ziesec@alfred.edu

19–23 October
Advanced Solid State Lasers Conference, Prague, Czech Republic
Optica Customer Service. Email custserv@optica.org Web www.optica.org

22 October
Health and safety workshop on the use of hydrogen in high temperature industries, Sheffield, UK
Web www.tickettailor.com/events/ukhyres/1825280

27–30 October
UNITECR 2025, Unified International Technical Conference on Refractories, Cancun, Mexico.
Email contact@unitecr2025.com Web unitecr2025.com

4–6 November
GlassBuild America, Orlando, Florida, USA.
National Glass Association, USA. Email attend@glass.org Web www.glass.org

6–9 November
Glasstech Asia and Fenestration Asia 2025, Jakarta, Indonesia.
Glasstech Asia Team. Email bauasia@mmiasia.com.sg and annie@mmiasia.com.sg Web glasstechasia.com.sg

10–12 November
M3P3: Materials, Minerals & Mining for People, Places, Planet, Birmingham, UK.
IoM3, 297 Euston Road, London NW1 3AD, UK. Web www.iom3.org

12–14 November
The Czech and Slovak Glass conference 2025, Trencin, Slovak Republic.
Slovak Glass Society, Sklárska 117, 020 61 Lednické Rovne, Slovak Republic. Email saczk@rona.glass Web rona.glass/SaCZK

17–20 November
11th Intl. Symp. on Sustainable Materials Recycling Processes & Products, Lapu-Lapu, Philippines.
Web www.flogen.org

20 November
Glass Focus Awards, Manchester, UK.
British Glass, 9 Churchill Way, Chapeltown, Sheffield, South Yorkshire S35 2PY, UK. Email glassfocus@britglass.co.uk Web www.britglass.co.uk

26–28 November
UZ Glass, Tashkent, Uzbekistan.
Email info@uzglass.com Web uzglass.com/

28–29 November
Glass History and Glass Design, Tübingen, Germany.
A. Gonzalez, DGG, Siemensstraße 45, 63071 Offenbach am Main, Germany. Email gonzalez@hvg-dgg.de Web www.hvg-dgg.de

2026

17–22 January
SPIE Photonics West, San Francisco, California, USA.
Email customerservice@spie.org Web spie.org

4–5 February
Glassman Italy 2026, Bologna, Italy.
Ken Clark, Glassman Events. Email kenclark@quartzltd.com Web www.glassmanevents.com

15–19 March
OFC 2026, Los Angeles, USA.
Email custserv@optica.org Web www.ofcconference.org

7–10 April
ChinaGlass 2026, Shanghai, China.
Email:ccscg@chinaglass-expo.com Web www.chinaglass-expo.com/

12–16 April
SPIE Photonics Europe, Strasbourg, France.
Email customerservice@spie.org Web spie.org

12–16 April
ACerS Spring Meeting (including the Glass and Optical Materials Division Annual Meeting), Bellevue, WA, United States.
The American Ceramic Society, 550 Polaris Parkway, Suite 510, Westerville, OH 43082, USA. Email customerservice@ceramics.org Web www.ceramics.org

13–17 April
ICG Annual meeting/18th ESG Conference, Lyon, France.
USTV, 114 Rue la Boétie, 75008 Paris, France. Email ustv@ustverre.fr Web www.ustverre.fr

19–23 April
ICG Spring School 2026, Lloret del Mar, Spain.
Dr Sophie Schuller Email sophie.schuller@cea.fr
Dr Thibault Charpentier Email thibault.charpentier@cea.fr

25–30 April
69th Annual SVC Technical Conference, Long Beach, California, USA.
Society of Vacuum Coaters (SVC), PO Box 10628, Albuquerque, NM 87184, USA. Email svcinfo@svc.org Web www.svc.org

27 April–1 May
Fourth São Carlos School on Glasses and Glass-ceramics, São Paulo and São Carlos, Brazil.
Email dedz@ufscar.br, certevlamav@gmail.com, lancelotti.r@dema.ufscar.br Web www.certev.ufscar.br/en

20–21 May
Glassman Latin America, Mexico City, Mexico.
Ken Clark, Glassman Events. Email kenclark@quartzltd.com Web www.glassmanevents.com

1–3 June
99th Annual Meeting of the DGG, Würtzburg, Germany.
Angela Gonzalez, Deutsche Glastechnische Gesellschaft (DGG), Siemensstraße 45, 63071 Offenbach am Main, Germany. Email gonzalez@hvg-dgg.de Web www.hvg-dgg.de

3–4 June
Furnace Solutions. St Helens, UK.
Christine Brown, Society of Glass Technology, 9 Churchill Way, Chapeltown, Sheffield S35 2PY, UK. Email christine@sgt.org Web www.sgt.org

3–6 June
Glass Art Society Annual Conference, Corning, USA.
Glass Art Society, 700 NW 42nd St #101, Seattle, WA 98107, USA. Email info@glassart.org Web www.glassart.org

16–17 June
Challenging Glass Conference 10 (International Conference on the Architectural and Structural Application of Glass), Delft, The Netherlands.
Email cgc10@challengingglass.com Web challengingglass.com

6–10 July
11th International Conference on Borate Glasses, Crystals

and Melts and Third International Conference on Phosphate Materials, Lille, France.
Gérald Lelong, Sorbonne Université - IMPMC, France. Email boratephosphate@sciencesconf.org Web boratephosphate.sciencesconf.org

27–30 July
Optica Advanced Photonics Congress, Long Beach, USA
Email custserv@optica.org Web www.ofcconference.org

27–31 August
International Festival of Glass 2026, St Helens, UK.
Glass Art Society, 700 NW 42nd St #101, Seattle, WA 98107, USA. Email info@glassart.org Web www.glassart.org

31 August–3 September
Sintering 2026, the International Conference on Sintering, Aachen, Germany.
Olivier Guillon, Forschungszentrum Jülich/RWTH Aachen University, Germany. Email sintering2026@fz-juelich.de Web www.sintering2026.org

6–11 September
ICC11, 11th International Congress on Ceramics, Sapporo, Japan.
ICC11 Office, The Ceramic Society of Japan, 2-22-17 Hyakunincho, Shinjuku, Tokyo 169-0073, Japan. Email ICC11@ceramic.or.jp Web www.ceramic.or.jp/en/icc11.html

7–9 September
SGT Annual Conference, Cambridge, UK.
Christine Brown, Society of Glass Technology, 9 Churchill Way, Chapeltown, Sheffield S35 2PY, UK. Email christine@sgt.org Web www.sgt.org

17–22 September
ICOOPMA 2026. 11th International Conference on Optical, Optoelectronic and Photonic Materials and Applications, Hamamatsu, Japan.
Web wwp.shizuoka.ac.jp/icoopma2026/

23–25 September
GlassBuild America, Las Vegas, Navada, USA.
National Glass Association NGA, 8200 Greensboro Drive, Suite 302, VA 22102 McLean, USA. Email attend@glass.org Web www.glass.org

4–7 October
ACerS 128th Annual Meeting with Materials Science and Technology, Pittsburgh, PA, USA.
The American Ceramic Society, 550 Polaris Parkway, Suite 510, Westerville, OH 43082, USA. Email customerservice@ceramics.org Web www.ceramics.org

20–23 October
Glasstec 2026, Düsseldorf, Germany.
Anke Seltmann, glasstec, D-40474 Düsseldorf, Am Staad (Stockumer Höfe), Germany. Email seltmanna@messe-duesseldorf.de Web www.glasstec-online.com/

16–22 November
MATÉRIAUX 2026, Grenoble, France.
Benoît Coasne, Président du comité local d'organisation Email benoit.coasne@univ-grenoble-alpes.fr Web materiaux2026.org

2027

1–6 May
70th Annual SVC Technical Conference, Dallas, Texas, USA.
Society of Vacuum Coaters (SVC), PO Box 10628, Albuquerque, NM 87184, USA. Email svcinfo@svc.org Web www.svc.org

18–20 May
FIT Show 2027, Birmingham, UK.
FIT Show, Montgomery Group, 9 Manchester Square, London W1U 3PL, UK. Email team@event.fitshow.co.uk Web www.fitshow.co.uk

23–27 July
16th International Conference on the Structure on Non-Crystalline Materials, Boulder, Colorado, USA.
Organisers: Jincheng Du (U North Texas), Sabyasachi Sen (UC Davis)

12–16 September
ICG 2027 Annual Meeting, Ilhabela, Brazil.

24–27 October
ACerS 129th Annual Meeting with Materials Science and Technology, Pittsburgh, PA, USA.
The American Ceramic Society, , USA. Email customerservice@ceramics.org Web www.ceramics.org

26–28 October
GlassBuild America, Orlando, Florida, USA.
National Glass Association, USA. Email attend@glass.org Web www.glass.org

14th International Symposium on Crystallization in Glasses and Liquids, Guangzhou, China.
Organiser: Shifeng Zhou, South China University of Technology, Guangzhou, China.

Ninth International Workshop on Flow and Fracture of Advanced Glasses, France.

ICG Annual Meeting 2027, Ilhabela, Brazil.
Co-chairs: Ana Candida M. Rodrigues, Edgar D. Zanotto and Hellmut Eckert

2028
3–9 April
Bauma 2028, Munich, Germany.
Bauma, Messe München, Germany. Email info@bauma.de Web bauma.de

6–11 May
71st Annual SVC Technical Conference, Nashville, Tennessee, USA.
Society of Vacuum Coaters (SVC), PO Box 10628, Albuquerque, NM 87184, USA. Email svcinfo@svc.org Web www.svc.org

23–27 July
ICC12, Int. Congress on Ceramics, Birmingham, UK.
Professor Jon Binner, School of Metallurgy and Materials, University of Birmingham, Edgbaston, Birmingham B15 2TT, UK. Email j.binner@bham.ac.uk

24–29 September
28th International Congress on Glass, Fukuoka, Japan.

10–12 October
GlassBuild America, Houston, Texas, USA.
National Glass Association NGA, 8200 Greensboro Drive, Suite 302, VA 22102 McLean, USA. Email attend@glass.org Web www.glass.org

22–25 October
ACerS 130th Annual Meeting with Materials Science and Technology, Pittsburgh, PA, USA.
The American Ceramic Society, 550 Polaris Parkway, Suite 510, Westerville, OH 43082, USA. Email customerservice@ceramics.org Web www.ceramics.org

2029
8–13 April
18th Pacific Rim Conference on Ceramic and Glass Technology including ACerS Annual Spring Meeting 2029, Waikoloa Village, HI, USA.
The American Ceramic Society, 550 Polaris Parkway, Suite 510, Westerville, OH 43082, USA. Email customerservice@ceramics.org Web www.ceramics.org

www.ingramcontent.com/pod-product-compliance
Lightning Source LLC
LaVergne TN
LVHW081553011026
849160LV00011BA/979